Making Sense of Statistics

Practice Questions

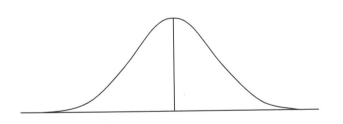

Keeok Park

© 2013 by Keeok Park ISBN: 978-1481179898

Contents

III. Sample Exam Questions 118

I. Practice Questions

Chapter 1

Hypothesis

1. David has the following research hypothesis: employees with a high level of emotional intelligence have fewer conflicts with other employees in the workplace than do individuals with a low level of emotional intelligence.

A. State the null hypothesis.

B. State the cases.

C. State the independent variable.

D. State the dependent variable.

2. Employees who use more technology in the process of performing their jobs are more productive than are employees who use less technology in the process of performing their jobs.

A. State the null hypothesis.

B. State the cases.

C. State the independent variable.

D. State the dependent variable.

3. A medical researcher thinks that hospitals headed by a transformational leader provide a more satisfactory work environment than hospitals headed by a transactional leader.

A. State the research hypothesis.

B. State the null hypothesis.

C. Identify the independent variable in the research hypothesis.

D. Identify the dependent variable in the research hypothesis.

4. Irene has the following hypothesis: employees who work in a team-based work environment have a higher sense of belonging than do employees who work in a non team-based work environment.

A. Is this a research hypothesis or a null hypothesis?

B. State the cases.

C. State the independent variable.

D. State the dependent variable.

5. Develop a research hypothesis that deals with the issues that you encounter in your workplace. State the research hypothesis, null hypothesis, cases, independent variable, and dependent variable.

6. Jake suspects that individuals with an optimistic world view make a longer–term investment than individuals with a pessimistic world view. State the research hypothesis, null hypothesis, cases, independent variable, and dependent variable.

7. John encounters the following statement from a newspaper article: Home loan applicants with high credit scores are more likely to exercise a strategic default than home loan applicants with low credit scores. State the research hypothesis, null hypothesis, cases, independent variable, and dependent variable.

8. Consider the following research hypothesis: Employees who feel a sense of meaning in their work (a sense that their work benefits not just their organization but the larger world) are more satisfied with their work than are employees who do not feel a sense of meaning in their work. State the null hypothesis, cases, independent variable, and dependent variable.

9. A clothing store manager who uses discount coupons to promote new clothes thinks that customers who use a coupon at the store purchase more items than customers who do not use a coupon at the store. State the research hypothesis, state the null hypothesis, cases, independent variable, and dependent variable.

10. A market researcher thinks that restaurant customers who pay their meals with a credit card give a higher amount of tips than restaurant customers who pay their meals with cash. State the research hypothesis, state the null hypothesis, cases, independent variable, and dependent variable.

Chapter 2

Levels of Measurement

1. At what level (nominal, ordinal, interval-ratio) is each of the following variables measured?

A. Number of company stocks listed in the United States.

B. Type of mutual fund: equity fund, fixed-income fund, money market fund (Note: According to the TDAmeritrade website, there are more than 13,000 mutual funds available for purchase. This is more than the number of stocks listed in the United States, which is less than 9,000).

C. Phone numbers of iPhone screen protector suppliers.

D. Degree of popularity of Android apps (very popular, somewhat popular, neutral, somewhat unpopular, very unpopular).

E. Method of payment for vehicle purchase (cash, lease, finance).

F. Percentage of residents who use cell phones for news.

G. Rank order of the five most common last names in the United States listed on the U.S. Census Bureau website: Smith, Johnson, Williams, Brown, and Jones.

H. Rank order of the five largest companies in the United States by revenue in 2012 (as listed on the Fortune website): Exxon Mobil, Wal-Mart Stores, Chevron, Conoco Phillips, General Motors.

2. David has the following research hypothesis: individuals with a high level of emotional intelligence are more successful in maintaining work relationships with their coworkers than are individuals with a low level of emotional intelligence. He measures the level of emotional intelligence with a question that has response categories from 5 (very high) to 1 (very low) and measures levels of success in maintaining work relationships with their coworkers with a question that has response categories from 7 (very high) to 1 (very low).

A. State the independent variable.

B. State the dependent variable.

C. At what level is the independent variable measured?

D. At what level is the dependent variable measured?

3. Denise has a hypothesis: employees with a high level of self-esteem have more conflicts with their coworkers in the workplace than employees with a low level of self-esteem. She measures self-esteem levels with a 5 point scale question (very high, high, neither high nor low, low, very low) and measures conflicts by counting the number of conflicts employees have with their coworkers.

A. State the independent variable.

B. State the dependent variable.

C. At what level is the independent variable measured?

D. At what level is the dependent variable measured?

4. A medical researcher hypothesizes that hospitals headed by a transformational leader provide a more spiritual work environment than hospitals headed by a transactional leader. She measures the leadership styles of hospitals by a survey question that determines whether a particular hospital is headed by a transformational leader or by a transactional leader. Then she measures the degree of spiritual work environment by a spiritual work environment index that ranges from 1 to 100. The index is created by adding a series of scores that measure different aspects of spirituality in the workplace.

A. State the independent variable.

B. State the dependent variable.

C. At what level is the independent variable measured?

D. At what level is the dependent variable measured?

5. You think that companies led by a more visionary CEO overcome recessions better than companies with a less visionary CEO.

A. State the research hypothesis.

B. State the independent variable.

C. State the dependent variable.

D. Explain at what level the independent variable is measured.

E. Explain at what level the dependent variable is measured.

6. Technology companies that co-create their new consumer products with customers, suppliers, independent contractors, and other companies are more likely to be successful with them than technology companies that solo-create their new consumer products internally. State the independent and dependent variables. Explain the level at which you would measure the independent and dependent variables.

7. Companies that use cloud computing are more efficient in information technology operations than companies that use their own dedicated data centers. State the independent and dependent variables. Explain the level at which you would measure the independent and dependent variables.

8. Develop a research hypothesis that deals with the issues that you encounter in your workplace. State the independent and dependent variables and explain the level at which you would measure the independent and dependent variables.

9. Consider the following research hypothesis: Employees respect managers who privately criticize, but publically praise them more than managers who privately praise, but publically criticize them. State the independent and dependent variables and explain the level at which you would measure the independent and dependent variables.

10. A manager of a manufacturing company is considering adopting a "just in time" inventory system. She thinks that it is more profitable than a "just in case" inventory system. State the research hypothesis and the null hypothesis. State the independent and dependent variables and explain the level at which you would measure the independent and dependent variables.

Chapter 3

Frequencies and Graphs

1. Would you draw a bar chart or a histogram for the following variable? Why?

A. Number of public companies that missed the most recent quarterly financial statement file deadline.

B. Size of companies (small, medium, and large).

C. Average prices of smart phones charged by top ten smart phone companies.

2. Construct a frequency distribution for the following variable by filling in the parentheses. (Notice that job sectors is a nominal variable.)

Number of Employees in Different Job Categories at a Company
Executives 20
Managers 50
Professionals 100
Clerical 60

Frequency Distribution of Job Sectors

Job Categories	Frequency	Proportion	Cumulative Proportion
Executives	20	( )	( )
Managers	50	( )	( )
Professionals	100	( )	( )
Clerical	60	( )	( )

3. A toy store sold the following number of toys during a Christmas season. Construct a frequency distribution for the number of toys sold variable by filling in the parentheses.

Stardoll by Barbie 50

Furby 45

Spider-Man Web Shooting Figure 35

Twister Dance 25

All others 100

Frequency Distribution of the Number of Toys Sold by
Different Toy Types

Toy Type	Frequency	Proportion	Cumulative Proportion
Stardoll by Barbie	50	( )	( )
Furby	45	( )	( )
Spider-Man Shooting Figure	35	( )	( )
Twister Dance	25	( )	( )
All Others	100	( )	( )

4. Speculate and describe the most probable shape for each of the following distributions.

A. Accounting test scores measuring the accounting abilities of 100 entry level bookkeepers that include 2 experienced CPAs. Positively skewed.

B. Amount of annual retirement contributions of all 10,000 employees of a typical company. Retirement contributions are based on employees' salaries. Positively skewed.

C. Actuary test scores of 50 actuaries who were well prepared for the test except for two actuaries who partied all night and then some before taking the exam. Negatively skewed.

D. French speaking ability test scores of a tourist group consisting of 50 French majors and 50 non-French majors. Bimodal.

E. Prices of pizzas in all pizza restaurants in a city. Roughly symmetric.

F. Amount of profits of 100 companies during a recession in which virtually all companies struggled except for five oil companies that made tons of money because of high gasoline prices. Positively skewed.

5. Would you use a bar chart or a histogram for each of the following variables? Why?

A. Number of available jobs.

B. Types of employees (full time, part time, temporary, leased, job share, and employees with co employers).

C. Number of companies that provide health insurance benefits to employees.

6. Would you use a bar chart or a histogram for each of the following variables? Why?

A. Number of smart phones sold in a quarter.

B. Number of private text messages sent by employees during work hours.

C. Degree of supportive work environment (more, less).

D. Type of companies (those that allow telecommuting and those that do not allow telecommuting)

7. Describe the most probable shape of the distribution of the following variables.

A. Prices for a thirty minute massage session charged by each of the 50 massage parlors in a medium sized city.

B. Number of cell phone minutes used in a month by a group of 100 cell phone users in which there are 2 professional game players who play games on their cell phones almost 24 hours a day and seven days a week.

C. Frequencies of donation activities of 50 companies that donate money for community organizations quite often except for three companies that do not.

D. Results of a financial audit of 100 companies that all passed with flying colors without any problems.

E. Profitability scores of 30 largest restaurants in an economically active region and those of 30 largest restaurants in an economically dormant region.

F. Number of ethics violations of 10 companies. Every company had a different number of ethics violations.

8. What would be the shape of the distribution of the following variables?

A. Number of times 100 companies were profitable in the quarters of the last 5 years. There were only 5 companies that were consistently profitable from the beginning to the end.

B. Average prices of corn dogs sold by all 10 vendors in a small city.

C. Amount of bonuses given by 70 financial management companies that always give at least 10% bonuses to their employees except for three companies that do not give any because their employees' salaries are higher.

D. Prices of Apple smart phones and prices of cheap cell phones that are sold only at discount stores.

E. Stress test results of 100 banks that all passed with flying colors except for six banks that are determined to have failed the test.

9. According to a Yahoo Finance web site, the following are the 10 best job opportunities of the future. On that web site, "the professions are ranked based on the number of job openings projected between 2010 and 2020 as a percentage of the 2010 headcount in that specific field" (Samuel Weigley and Alexander E. M. Hess, 2012). Construct a frequency distribution for the job opportunities variable (specifically, the number of new job openings) by filling in the parentheses.

1. Actuaries
Future job openings as a percentage of 2010 employment: 87.1%
New Openings, 2010 to 2020: 18,900

2. Glaziers (glass installers)
Future job openings as a pct. of 2010 employment: 79.7%
New openings, 2010 to 2020: 33,400

3. Statisticians
Future job openings as a pct. of 2010 employment: 74.5%
New openings, 2010 to 2020: 18,700

4. Pest Control Workers
Future job openings as a pct. of 2010 employment: 70.9%
New openings, 2010 to 2020: 48,500

5. Interpreters and Translators
Future job openings as a pct. of 2010 employment: 69%
New openings, 2010 to 2020: 40,300

6. Optometrists
Future job openings as a pct. of 2010 employment: 68.4%
New openings, 2010 to 2020: 23,400

7. Natural Science Managers
Future job openings as a pct. of 2010 employment: 68%
New openings, 2010 to 2020: 33,500

8. Market Research Analysts and Marketing Specialists
Future job openings as a pct. of 2010 employment: 67.8%
New openings, 2010 to 2020: 191,800

9. Insulation Workers
Future job openings as a pct. of 2010 employment: 67.5%
New openings, 2010 to 2020: 34,700

10. Environmental Science and Protection Technicians, Including Health
Future job openings as a pct. of 2010 employment: 65.9%
New openings, 2010 to 2020: 19,500

Frequency Distribution of Top Ten New Job Openings

Job Opportunities	Frequency	Proportion	Cumulative Proportion
Actuaries	18,900	( )	( )
Glaziers	33,400	( )	( )
Statisticians	18,700	( )	( )
Pest Control Workers	48,500	( )	( )
Interpreters	40,300	( )	( )
Optometrists	23,400	( )	( )
Natural Science Managers	33,500	( )	( )
Marketing Analysts	191,800	( )	( )
Insulation Workers	34,700	( )	( )
Environmental Technicians	19,500	( )	( )

10. According to the Fortune Magazine website, the following are the top 5 best large companies to work for and their employees. Construct a frequency distribution for the number of employees variable.

1. Google 18,500
2. Boston Consulting Group 1,958
3. SAS Institute 6,046
4. Wegmans Food Markets 41,717
5. Edward Jones 36,937

Chapter 4

Central Tendency and Dispersion Measures

1. Find the median in each of the following data sets. The variable is the number of new products companies began to sell this year.

A. 6 7 8   B. 2 8 6 2

2. John has a variable: the number of overtime hours. He has 5 employees. His five employees incur the following number of overtime hours: 0 3 5 7 10

A. Calculate the mean.

B. Calculate and interpret the standard deviation.

3. Melissa is an analyst in a customer services department. She has the following means and standard deviations regarding her customers' ratings on the quality of the services they received from 100 customer service representatives in two customer services units. The ratings are an index that ranges from 1 (lowest quality) to 100 (highest quality). Interpret each of the two standard deviations. Then compare the two standard deviations.

SweepingFeet Unit: Mean = 80.0   Standard Deviation = 0   N = 50
GoingExtraMiles Unit: Mean = 80.0   Standard Deviation = 20   N = 50

4. Consider the following data. The values (scores) represent the number of times five employees in an office (Beach Office) had lunch with their coworkers in a month.   3, 3, 4, 5, 5

A. Calculate and interpret the standard deviation.

B. Consider the following data from another office (Coastal Office). Each value represents the number of times each of the five employees had lunch with their coworkers in a month. Calculate and interpret the standard deviation.   0, 0, 4, 8, 8

C. Now compare the two standard deviations in A and B. What can we say about the lunching behaviors of the employees in A and B? If all other factors are equal, would you prefer to work in Beach Office or in Coastal Office? Why?

5. The following central tendency and dispersion measures are about the amounts of salaries of the employees of two corporations: White Gold and Yellow Gold.

White Gold
Mean = $80,000
Median = $80,000
Standard deviation = $10,000

Yellow Gold
Mean = $80,000
Median = $40,000
Standard deviation = $70,000

A. Which salary distribution of the two corporations is more symmetric (and perhaps more equitable)?

B. Interpret each of the two standard deviations.

C. Compare the two standard deviations. What can we say about the salaries of the employees of the two corporations?

6. Come up with some hypothetical or real numbers for a variable that is related to your workplace. Calculate and interpret the mean and the standard deviation of those numbers.

7. The following are customer satisfaction ratings of a few major smart phones produced by different manufacturers reported by J.D. Power and Associates on its website. The scores are customer satisfaction indexes that are obtained from four different aspects of their wireless devices: performance, physical design, features, and ease of operation. It is based on a 1000 point scale.

Apple 849
HTC 790
Samsung 782
Motorola 777
Nokia 763
LG 742
RIM BlackBerry 740
HP/Palm 707

A. Find the median.

B. Calculate the mean.

C. Calculate and interpret the standard deviation.

8. The following are the means and standard deviations of productivity measures of sales employees in two different divisions of a company. Productivity measures range from 1 (lowest) to 100 (highest). What can we say about the productivity behavior of sales employees in these two divisions?

	Marathon Division	Non Stop Division
Mean	80	80
Standard Deviation	2	10
	N = 100	N = 100

9. Dividend yield is obtained by dividing the amount of annual dividend per share by the current share price of a company's stock (expressed as a percentage). The formula is: (annual dividend per share/current stock price per share) × 100. The following are ten DOW companies with the highest dividend yield reported in 2012 on the IndexArb website.

Stock	Price	Dividend	Yield (%)
AT&T	33.42	1.8000	5.39
Verizon Communications	41.70	2.0750	4.98
Intel	20.03	0.9300	4.64
Hewlett Packard	13.08	0.5540	4.24
Merck	42.80	1.7600	4.11
Dupont	42.10	1.7200	4.09
Pfizer	23.66	0.9600	4.06
Chevron	101.62	3.8700	3.81
General Electric	20.06	0.7600	3.79
McDonalds	84.05	3.0800	3.66

A. Find the median of the dividend yield.

B. Calculate the mean.

C. Calculate and interpret the standard deviation.

10. You are a manager of a technology company who is interested in using the concept of return on investment (ROI) in planning for the future of the company. ROI is obtained by dividing the net profit (gain from investment – cost of investment) by the cost of investment. It is usually expressed as a percentage. Your company has two product development divisions and they develop various technology products. One division focuses on developing tablets and another division focuses on smart phones. The following are their ROI numbers (%) from seven major products that each division developed in the last 2 years.

	Smart Phone Division	Tablet Division
Product #1	0	100
Product #2	0	100
Product #3	0	0
Product #4	300	200
Product #5	0	100
Product #6	0	200
Product #7	400	0

A. Calculate and interpret the standard deviation for each of the two divisions.

B. Compare the two standard deviations. What can we say about the performance of the two divisions?

C. Discuss some managerial implications.

Chapter 5

The Normal Distribution

1. The manager of Upscale Burger Corporation learns that her 50 stores sold an average of 300 burgers with a standard deviation of 100 burgers on a spring day. The numbers of burgers sold are normally distributed. What is the probability of finding a store that sold more than 500 burgers?

2. A marketing researcher learns that young adults in his city purchased an average of 30 songs online with a standard deviation of 10 songs last year. She obtained this information from a survey of 500 young adults. Assume that the numbers of songs purchased are normally distributed. What is the probability of a young adult purchasing less than 10 songs last year?

3. A claims adjustment manager learns that her 200 employees (N = 200) spend an average of 5 hours per claims case with a standard deviation of 2 hours. The numbers of hours are normally distributed. What is the probability of finding an employee who spends more than 11 hours per claims case?

4. An educator learns that teenagers in her school spend an average of 3 hours a day surfing the Internet with a standard deviation of 2 hours. The numbers of hours are normally distributed. She has 500 teenagers in her school.

A. What is the probability of finding a teenager who spends more than 7 hours a day surfing the Internet?

B. What is the probability of finding a teenager who spends less than 1 hour a day surfing the Internet?

C. How many teenagers spend more than 9 hours a day surfing the Internet?

5. Cindy Lou learns that in the city of Hills, its 10,000 residents are planning to spend an average of 1,000 dollars on Christmas gifts this year. The standard deviation is 300 dollars. Assume that the amounts are normally distributed.

A.   What is the probability of spending more than 1,600 dollars?

B.   What is the probability of spending less than 100 dollars?

C.   What percentage of the residents will spend more than 1,900 dollars?

D.   How many residents will spend less than 400 dollars?

6. A recent study finds that in California, 24% of the drivers involved in fatal crashes had invalid (expired, revoked, suspended, or no) driver licenses. The average invalid license rate for all states is 18%, with a standard deviation of 4%. Assuming that the rates are normally distributed across the 50 states, find the number of states that have a higher invalid license rate than California. (Hint: first, we calculate the z score that goes with 24%. Second, we go to the standard normal distribution table (Table 5.1) to get the probability that is associated with that z score. The standard normal distribution table will give us the probability of an area that is beyond 24% in the normal distribution with a mean of 18% and a standard deviation of 4%. Third, we calculate the number of states by multiplying that probability by the total number of cases (50).

7. A bank credit analyst learns that the average credit score (FICO score) is 700 with a standard deviation of 40 in a group of 1000 bank customers. Assume that individuals' credit scores are normally distributed.

A. Susie's credit score is 820. What is the probability of finding a person whose credit score is higher than Susie's?

B. How many individuals' credit scores are higher than Susie's?

C. Jason's credit score is 620. What is the probability of finding a person whose credit score is lower than Jason's?

D. How many individuals' credit scores are lower than Jason's?

E. The bank manager tells the credit analyst that the top 10% of the 1000 customers in terms of their credit scores are eligible for a preferred mortgage rate. Melissa's credit score is 780. Is she eligible for a preferred mortgage rate?

8. The U.S. News and World Report website lists America's most expensive hotel rooms. The following are the prices of the top ten most expensive ones. Assume that the prices are normally distributed and that no discount will be given.

1. Palms Casino Resort, Las Vegas Cost: $40,000 per night
2. Four Seasons Hotel New York, New York City Cost: $40,000 per night
3. The Plaza Hotel, New York City Cost: $30,000 per night
4. The St. Regis New York, New York City Cost: $21,000 per night
5. New York Palace, New York City Cost: $19,000 per night
6. Mandarin Oriental, New York City Cost: $18,000 per night
7. Beverly Hills Hotel and Bungalows, Los Angeles Cost: $17,300 per night
8. Four Seasons Hotel, Washington D.C. Cost: $15,000 per night
9. The Fairmont San Francisco, San Francisco Cost: $15,000 per night
10. The Joule, A Luxury Collection Hotel, Dallas Cost: $8,000 per night

A. Calculate the mean.

B. Calculate and interpret the standard deviation.

C. Calculate and interpret the z score for Palms Casino Resort.

D. Calculate and interpret the z score for the Fairmont San Francisco.

E. What is the probability of finding a room that charges more than $21,000?

F. What is the probability of finding a room that charges less than $15,000 in this distribution?

9. The following are the average prices of small (7 inch) tablet computers. Assume that the prices are normally distributed.

Google Nexus $200
Apple iPad Mini $320
Amazon Kindle Fire HD $200
Barnes and Noble Nook HD $200
Samsung Galaxy Tab $200
Amazon Kindle Fire $180
Polaroid Tablet $100
Blackberry Playbook $180
Koby Kyros $70
Pandigital $60

A. Calculate the mean.

B. Calculate and interpret the standard deviation

C. Calculate and interpret the z score for Samsung Galaxy Tab.

D. Calculate and interpret the z score for Polaroid Tablet.

E. What is the probability of finding a tablet that costs more than $390?

F. What is the probability of finding a tablet that costs less than $25 in this distribution?

10. A newspaper reports that the average household credit card debt in the United States is $7,000 with a standard deviation of $3000. There are 120 million households in the United States. Assume that household credit card debts are normally distributed.

A. What is the probability that the credit card debt for a household is more than $16,000?

B. What is the probability that the credit card debt for a household is less than $1,000?

C. How many households have a credit card debt of more than $10,000?

Chapter 6

Optional Chapter: Sampling, Sampling Distribution, and Central Limit Theorem

1. A manager of ForeverRun, a tire manufacturing company, claims that a brand of their radial tires, Superradial, provides at least 70,000 miles of wear. Her research assistant examines 100 tires and finds that the average life of tires is 71,000 miles with a standard deviation of 1000 miles. Construct and interpret a 95% confidence interval around the mean of 71,000 miles of wear.

2. In Question 1, an analyst wants to make sure that the average life of tires is at least 70,000 miles. Do you think he will prefer to use a smaller sample or a larger sample in the process of drawing that conclusion?

3. A manager of a smart phone manufacturing company is interested in establishing the average life of batteries that his company uses for their smart phones. His assistant tests 100 batteries and finds that the average life of batteries is 12 hours with a standard deviation of 2 hours. Construct and interpret a 95% confidence interval around the mean of 12 hours.

4. In Question 3, a competitor claims that their batteries last at least 13 hours. Can the manager in Question 3 claim that his batteries last as long as 13 hours?

5. A human resources manager of a large company is interested in knowing about the average salaries of clerical employees of her company. Her assistant gathers salary information of a sample of 50 clerical employees and finds that the average salary is $50,000 with a standard deviation of $10,000. Construct and interpret a 95% confidence interval around the sample mean of $50,000.

6. In Question 5, a rival company claims that the salaries of their clerical employees are $52,000 on average. That company spreads rumors about their "superior" salaries and tries to lure away employees. Based on your knowledge about the nature of confidence interval, what can you do to respond to that rumor?

7. An automobile analyst at J.D. Power & Associates wants to find out whether Honda customers' vehicle buying satisfaction scores are improving. She asks a sample of 200 new Honda buyers to evaluate their vehicle buying experience with a particular dealership. The items asked include reasonableness of price, friendliness of selling agents, smoothness of the buying process, and other items. The satisfaction index for the dealership ranges from 1 to 1000. Assuming that she received the following figures from the survey, construct and interpret a 95 percent confidence interval around the mean satisfaction score.

Mean Satisfaction Score	Standard Deviation	N
900	50	200

8. In Question 7, the automobile analyst at J.D. Power & Associates encounters a buying satisfaction score of 820 that was obtained for a rival automobile company. Can she claim that that score is lower than Honda's?

9. A manger of an online retailer wants to make sure that her company does not undersell its competitors. She institutes a pricing policy called "dynamic pricing" by which the competitors' prices are matched to make sure that they are comparable. Her assistants check the prices of a popular computer tablet sold by a sample of 15 retailers several times and report that the average price is $199 with a standard deviation of $10. The company's current price is $208. If the company's current price is not within the 95% confidence interval of the average price of $199, the manager plans to adjust the company's current price. Construct and interpret a 95 percent confidence interval around the mean price of $199.

10. In Question 9, a rival online retailer claims that their average price in a one week period was $197 and therefore, their price is the lowest among all online retailers on average. Can this claim be justified based on a 95% confidence interval?

Chapter 7

Population Mean Compatibility Test: Hypothesis Testing with One Sample Mean

1. Employees of an engineering company claim that they are paid less than the average salary of all engineering employees in the state. As a Union representative, you agree with them and want to test that claim. You find out that the average of salary of all engineering employees in in the state is $70,000. You survey a sample of 100 employees in your company and learn that the average salary is $65,000 with a standard deviation of $7,000. Do a hypothesis test.

1. What is the research hypothesis?

2. What is the null hypothesis?

3. Should we do a one-tailed test or a two-tailed test?

4. Should we do a z test or a t test?

5. What is the value of the standard error?

6. What is the value of the test statistic?

7. What is the rejection region?

8. What is your decision?

9. What is the conclusion?

2. This year, the average monthly charge for unlimited call, text, and data plans by four major cell phone service providers is $100, excluding taxes. As a young and restless researcher for a cell phone users group, you think average cell phone users pay more than that in your region because of extra charges and fees. You survey 90 unlimited plan users and find that they pay $105 on average with a standard deviation of $10. Can you confidently say that they are paying more than the average monthly charge of $100? Answer the following questions.

1. What is the research hypothesis?

2. What is the null hypothesis?

3. Should we do a one-tailed test or a two-tailed test? Why?

4. Should we do a z-test or a t-test? Why?

5. What is the value of the standard error?

6. What is the value of the test statistic?

7. What is the rejection region?

8. What is your decision?

9. What is the conclusion?

3. Jack is a publicist working for a pop singer who is as popular as Taylor Swift. He needs to write a market analysis report to his client regarding the sale of her new album. Last month the music stores in the Los Angeles area sold an average of 300 copies of her new album both online and in store. At the end of the current month, he quickly surveys a random sample of 50 stores to estimate the average number of copies sold per store this month. Can he conclude with confidence that the average number of store sales is higher this month?

Number of copies sold	Standard deviation	Number of stores
310	30	50

1. What is the research hypothesis?

2. What is the null hypothesis?

3. Should we do a one-tailed test or a two-tailed test? Why?

4. Should we do a z-test or a t-test? Why?

5. What is the value of the standard error?

6. What is the value of the test statistic?

7. What is the rejection region?

8. What is the decision?

9. What is the conclusion?

4. The Office of Real Estate Assessment of Suffolk County, New York purchased a new software program called ErrorFree to reduce the number of errors in its real estate assessment. Last year, the average number of appeals for the office's real estate assessment notices was 100 per district. This year, the office manager randomly selects 18 districts and learns that the number of appeals per district is 95 with a standard deviation of 30. Assuming that other factors remained the same from last year to this year, did the new software program reduce the average number of appeals per district?

1. What is the research hypothesis?

2. What is the null hypothesis?

3. Should we do a one-tailed test or a two-tailed test? Why?

4. Should we do a z-test or a t-test? Why?

5. What is the value of the standard error?

6. What is the value of the test statistic?

7. What is the rejection region?

8. What is the decision?

9. What is the conclusion?

5. You are a health administrator in a city. You think that the amount of time senior citizens take to recover from the common cold is different from the amount of time the general public takes to recover from the common cold. A previous study reports that the average recovery time for the general public was 10 days. You take a random sample of 100 senior citizens and find that the average recovery time for senior citizens is 11 days with a standard deviation of 2 days.

1. What is the research hypothesis?

2. What is the null hypothesis?

3. Should we do a one-tailed test or a two-tailed test? Why?

4. Should we do a z-test or a t-test? Why?

5. What is the value of the standard error?

6. What is the value of the test statistic?

7. What is the rejection region?

8. What is the decision?

9. What is the conclusion?

6. Suppose the employees of a large technology company claim that they are paid less than the state average. You agree with them and want to test that claim. The average salary of all employees of technology companies in the state is $90,000. The average salary of a sample of 100 employees of the large technology company is $89,000 with a standard deviation of $9,000. Do a hypothesis test by answering the following 9 sub questions.

1. What is the research hypothesis?

2. What is the null hypothesis?

3. Should we do a one-tailed test or a two-tailed test? Why?

4. Should we do a z-test or a t-test? Why?

5. What is the value of the standard error?

6. What is the test statistic?

7. What is the rejection region?

8. What is the decision?

9. What is the conclusion?

7. Several vendors who supply equipment to a large telecommunications company claim that the company takes more than 60 days to process their invoices. In the past, the company promised that it would pay all invoices within 60 days from receipt. An accountant wants to test the vendors' claim without auditing all the invoices. She randomly pulls out 150 invoices received in the last 6 months and examines the number of days the company took to process them. She learns that the average number of processing days was 70 with a standard deviation of 10. Is the vendors' claim justified? Answer the following questions.

1. What is the research hypothesis?

2. What is the null hypothesis?

3. Should we do a one-tailed test or a two-tailed test? Why?

4. Should we do a z-test or a t-test? Why?

5. What is the value of the standard error?

6. What is the test statistic?

7. What is the rejection region?

8. What is the decision?

9. What is the conclusion?

8. Last year the Human Resources Department of a large transportation company implemented a diversity-training program in which employees learned about the value of a diverse workforce, including unique contributions from different genders, ethnic groups, nationalities, and others to the mission of the department. One of the reasons for the implementation of the program was that the number of conflicts among individuals with different backgrounds in the department increased in the last few years. The HR Director randomly selected 60 work units from all work units and examined the number of diversity-related conflicts in a 12-month period. The average number of conflicts of the 60 work units turned out to be 15 with a standard deviation of 5. If the mean number of conflicts last year was 20, can we say that the program actually reduced the average number of conflicts?

1. What is the research hypothesis?

2. What is the null hypothesis?

3. Should we do a one-tailed test or a two-tailed test? Why?

4. Should we do a z-test or a t-test? Why?

5. What is the value of the standard error?

6. What is the value of the test statistic?

7. What is the rejection region?

8. What is the decision?

9. What is the conclusion?

9. The manager of a large financial services company wants to find out whether switching the health insurance policy for employees to a new HMO will save money for the company. Last year the average monthly health insurance premium for the company was $500 per employee. This year, the manager implemented a voluntary pilot HMO program, which the manager is considering selecting for the entire workforce.

After 12 months of implementation of the pilot program, the manager wants to determine whether the new program is working without a full audit, which is costly. She randomly selects 40 employees from 500 employees who are enrolled in the pilot program and finds that the average monthly premium of the city for these 40 employees is $490 with a standard deviation of $50. Is the average monthly premium for the company in the pilot HMO program lower than the average premium that the company paid for its employees last year? Answer the following questions.

1. What is the research hypothesis?

2. What is the null hypothesis?

3. Should we do a one-tailed test or a two-tailed test? Why?

4. Should we do a z-test or a t-test? Why?

5. What is the value of the standard error?

6. What is the value of the test statistic?

7. What is the rejection region?

8. What is the decision?

9. What is the conclusion?

10. The manager of a Rich branch of the Loyalty Investment Company thinks that her account managers are more productive than the account managers of all the other branches in the state. She defines productivity as the amount of new money each account manager brings in. She selects a random sample of 45 account managers and finds that the average amount of new assets added to their accounts this year is $1million with a standard deviation of $0.5 million. The average amount of new assets added this year to the accounts of the managers of all the other branches in the state is $1.2 million. Can she claim that her account managers are more productive than the account managers of all the other branches in the state?

1. What is the research hypothesis?

2. What is the null hypothesis?

3. Should we do a one-tailed test or a two-tailed test? Why?

4. Should we do a z-test or a t-test? Why?

5. What is the value of the standard error?

6. What is the value of the test statistic?

7. What is the rejection region?

8. What is the decision?

9. What is the conclusion?

Chapter 8

Difference of Means Testing: Hypothesis Testing with Two Sample Means

1. A technology company manager is exploring the possibility of manufacturing smart phones in another country. She learns that smart phone manufacturing cost includes the following 12 major components: Memory, Housing/mechanicals, Display, Interface, Baseband processor, Applications processor, Camera, Radio frequency, GPS/WiFi/Bluetooth, Assembly, Battery, Power management. She suspects that the manufacturing costs of smart phones by international companies are lower than those by U.S. companies. By contacting manufacturers, consultants, inside sources, and others, she comes up with the following sample average costs per device and their standard deviations. Do a difference of means test.

	Average manufacturing cost	Standard deviation	N
International companies	$200	$40	52
U.S. companies	$250	$50	50

1. What is the research hypothesis?

2. What is the null hypothesis?

3. What is the independent variable?

4. What is the dependent variable?

5. At what level is the independent variable measured?

6. At what level is the dependent variable measured?

7. Should we do a one-tailed test or a two-tailed test? Why?

8. Should we do a z-test or a t-test? Why?

9. What is the value of the standard error?

10. What is the rejection region?

11. What is the value of the test statistic?

12. What is the decision?

13. What is the conclusion?

2. A brand name sunglasses chain owner is considering using low-cost leadership as its marketing strategy, adopting a concept from Porter (1984)'s generic strategies. He surveys the prices of polarized sunglasses in his own stores and those in his rival stores. The following are the average prices (in dollars) of polarized sunglasses that he thinks should be used for low-cost leadership. They are obtained from a sample of 100 stores altogether. He is interested in finding out whether his store's average price is lower than that of his rivals.

	Mean	Standard deviation	Number of cases
Own stores	30.00	10	50
Rival stores	35.00	15	50

1. What is the research hypothesis?

2. What is the null hypothesis?

3. What is the independent variable?

4. What is the dependent variable?

5. At what level is the independent variable measured?

6. At what level is the dependent variable measured?

7. Should we do a one-tailed test or a two-tailed test? Why?

8. Should we do a z-test or a t-test? Why?

9. What is the value of the standard error?

10. What is the rejection region?

11. What is the value of the test statistic?

12. What is the decision?

13. What is the conclusion?

3. A clothing marketing manager is interested in using "nostalgia marketing," a strategy that tries to rekindle a yearning for the past and a sense that previous times were better. She expects that stores that adopt a nostalgia marketing (NM) strategy have a higher amount of sales increase (in dollars) than stores that do not adopt a nostalgia marketing strategy. After implementing the strategy selectively in her clothing stores, she obtains the following sales information from a sample of 120 stores. The increase in daily sales was obtained by subtracting the amount of daily sales before the marketing strategy was adopted from the amount of daily sales after the marketing strategy was adopted.

	Average increase in daily sales	Standard deviation	N
Stores with NM	2,000	500	60
Stores without NM	1,000	500	60

1. What is the research hypothesis?

2. What is the null hypothesis?

3. What is the independent variable?

4. What is the dependent variable?

5. At what level is the independent variable measured?

6. At what level is the dependent variable measured?

7. Should we do a one-tailed test or a two-tailed test? Why?

8. Should we do a z-test or a t-test? Why?

9. What is the value of the standard error?

10. What is the rejection region?

11. What is the value of the test statistic?

12. What is the decision?

13. What is the conclusion?

4.You are a forensic accountant who is interested in the effect of the Sarbanes–Oxley Act of 2002 on the number of accounting scandals. Accounting scandals refer to misdeeds of misusing funds, overstating revenues, understating expenses, inflating g corporate assets, and others. As a result of the Sarbanes–Oxley Act of 2002, top executives must personally certify the accuracy of financial information that their companies report to government agencies. Because of this requirement, you think that fewer executives and employees were willing to commit fraud after the passage of the law. In order to test this hypothesis, you select a sample of 500 public companies and document all accounting infractions and calculate monthly infraction averages for 5 years before the passage of the law and for 5 years after the passage of the law .

Average number of monthly accounting scandals before and after
the Sarbanes–Oxley Act of 2002

	Mean	Standard deviation	Number of cases
Before	10	5	60
After	3	1	60

1. What is the research hypothesis?

2. What is the null hypothesis?

3. What is the independent variable?

4. What is the dependent variable?

5. At what level is the independent variable measured?

6. At what level is the dependent variable measured?

7. Should we do a one-tailed test or a two-tailed test? Why?

8. Should we do a z-test or a t-test? Why?

9. What is the value of the standard error?

10. What is the rejection region?

11. What is the value of the test statistic?

12. What is the decision?

13. What is the conclusion?

5. A telecommunications analyst is interested in finding out who pay more for monthly cell phone bills. She thinks that individuals who use gaming applications on their service pay more than individuals who do not use gaming applications on their service. By surveying a sample of 300 individuals in a region, she obtains the following information. Do a difference of means test.

Average amount of payments in dollars by gaming application use status

	Mean	Standard deviation	Number of cases
Gaming application users	$70	20	120
Non gaming application users	$50	10	180

1. What is the research hypothesis?

2. What is the null hypothesis?

3. What is the independent variable?

4. What is the dependent variable?

5. At what level is the independent variable measured?

6. At what level is the dependent variable measured?

7. Should we do a one-tailed test or a two-tailed test? Why?

8. Should we do a z-test or a t-test? Why?

9. What is the value of the standard error?

10. What is the rejection region?

11. What is the value of the test statistic?

12. What is the decision?

13. What is the conclusion?

6. A telecommunications analyst is interested in finding out who pay more for monthly cell phone bills. She thinks that individuals who use social media application on their service pay more than individuals who do not use social media application on their service. By surveying a sample of 300 individuals in a region, she obtains the following information. Do a difference of means test.

Average amount of payments in dollars by social media application use status

	Mean	Standard deviation	Number of cases
Social media application users	$ 65	10	160
Non users	$50	10	140

1. What is the research hypothesis?

2. What is the null hypothesis?

3. What is the independent variable?

4. What is the dependent variable?

5. At what level is the independent variable measured?

6. At what level is the dependent variable measured?

7. Should we do a one-tailed test or a two-tailed test? Why?

8. Should we do a z-test or a t-test? Why?

9. What is the value of the standard error?

10. What is the rejection region?

11. What is the value of the test statistic?

12. What is the decision?

13. What is the conclusion?

7. A telecommunications analyst thinks that prepaid plan users pay less for unlimited talk, text, and data plans than postpaid plan users. By surveying a sample of 200 individuals in a region, she obtains the following information. Do a difference of means test.

Average amount of monthly payments in dollars by payment plan type

	Mean	Standard deviation	Number of cases
Prepaid plan users	$ 50	10	50
Postpaid plan users	$100	10	150

1. What is the research hypothesis?

2. What is the null hypothesis?

3. What is the independent variable?

4. What is the dependent variable?

5. At what level is the independent variable measured?

6. At what level is the dependent variable measured?

7. Should we do a one-tailed test or a two-tailed test? Why?

8. Should we do a z-test or a t-test? Why?

9. What is the value of the standard error?

10. What is the rejection region?

11. What is the value of the test statistic?

12. What is the decision?

13. What is the conclusion?

8. An automobile industry analyst thinks that the average amount of ownership costs of hybrid vehicles is different from those of gasoline powered vehicles. Because he does not know which are lower, he surveys a sample of vehicle owners in a city and asks them about their ownership costs including original purchase price, gasoline cost, maintenance expense, and others. He then derives five year ownership costs from their answers and eliminates any owners whose vehicles were worth more than $50,000 originally. Do a difference of means test.

Average amount of ownership costs in dollars by vehicle type

	Mean	Standard deviation	Number of cases
Hybrid vehicle owners	$38,000	8,000	50
Gas powered vehicle owners	$37,000	10,000	150

1. What is the research hypothesis?

2. What is the null hypothesis?

3. What is the independent variable?

4. What is the dependent variable?

5. At what level is the independent variable measured?

6. At what level is the dependent variable measured?

7. Should we do a one-tailed test or a two-tailed test? Why?

8. Should we do a z-test or a t-test? Why?

9. What is the value of the standard error?

10. What is the rejection region?

11. What is the value of the test statistic?

12. What is the decision?

13. What is the conclusion?

9. A clothing retail industry manager thinks that Black Friday promotions are an effective promotion tool that adds to the bottom line of the retail chains. To test his initial idea, he collects information about annual store sales per square foot from 100 comparable stores and classify them into those that heavily use Black Friday promotions and those that do not. Do a difference of means test.

Average annual sales per square foot by Black Friday promotion status

	Mean	Standard deviation	# of cases
Stores with heavy BF promotion	$320	80	60
Stores without heavy BF promotion	$300	90	40

1. What is the research hypothesis?

2. What is the null hypothesis?

3. What is the independent variable?

4. What is the dependent variable?

5. At what level is the independent variable measured?

6. At what level is the dependent variable measured?

7. Should we do a one-tailed test or a two-tailed test? Why?

8. Should we do a z-test or a t-test? Why?

9. What is the value of the standard error?

10. What is the rejection region?

11. What is the value of the test statistic?

12. What is the decision?

13. What is the conclusion?

10. A human resources manager in a large corporation thinks that employees who live beyond 10 miles from their work place are more likely to be absent from work than those who live within 10 miles from their work place. To test his initial idea, he collects information about the number of absences in a year from a sample of 100 employees. Do a difference of means test.

Average number of annual absences by residence

	Mean	Standard deviation	Number of cases
Residence beyond 10 miles	10	4	55
Residence within 10 miles	9	3	45

1. What is the research hypothesis?

2. What is the null hypothesis?

3. What is the independent variable?

4. What is the dependent variable?

5. At what level is the independent variable measured?

6. At what level is the dependent variable measured?

7. Should we do a one-tailed test or a two-tailed test? Why?

8. Should we do a z-test or a t-test? Why?

9. What is the value of the standard error?

10. What is the rejection region?

11. What is the value of the test statistic?

12. What is the decision?

13. What is the conclusion?

Chapter 9

Analysis of Variance (ANOVA)

1. A convention coordinator wants to know whether the mean prices of hotel rooms in three different cities are different. A colleague claims that the average hotel room prices of all three cities—Las Vegas, San Diego, and Palm Springs—are the same. She disagrees with the coordinator's thinking that the average price of at least one city is different from those of other cities. She surveys 20 hotels in each city and calculates average prices. Answer the following questions.

A. What is the research hypothesis?

B. What is the null hypothesis?

C. What is the independent variable?

D. What is the dependent variable?

E. At what level is the independent variable measured?

F. At what level is the dependent variable measured?

G. If we want to do an f-test, what more information do we need?

2. A mutual fund investor thinks that buying index mutual funds is the most effective way to invest in mutual funds. In order to test her idea, she selects a sample of 20 index funds, 20 load mutual funds, 20 no load mutual funds in the value mutual fund category. The average annual return in the last five years was 7.9%, 7.0%, and 8.3%, respectively, after deducting all load and other fees. Set up the research and null hypotheses.

3. A management analyst thinks that self-employed individuals are more satisfied with their jobs than individuals working for someone else in large organizations or those working for someone else in small organizations. To show his colleagues that his hypothesis has some support, he measures various individuals' job satisfaction with an index that has three components—satisfaction with supervisors, satisfaction with coworkers, and satisfaction with subordinates. Each of these components is measured by a question on a ten point scale. The job satisfaction index ranges from 3 (lowest job satisfaction) to 30 (highest job satisfaction). He surveys ninety individuals with 30 in each group. Answer the following questions.

A.   What is the research hypothesis?

B.   What is the null hypothesis?

C.   What is the independent variable?

D.   What is the dependent variable?

E.   At what level is the independent variable measured?

F.   At what level is the dependent variable measured?

G.   If we want to do an f-test, what more information do we need?

4. The president of a toy manufacturing company is interested in assessing the effectiveness of three different online marketing strategies: social networking, Google search, and banner ads. She thinks that the Google search strategy generates a higher amount of sales than the other strategies. After a thirty day trial, she collects the information about the sales that resulted from fifteen customers who purchased toys from the company because of one of the three marketing efforts. Do an f test.

Sales ($) by Three Online Marketing Strategies

Social networking	Google Search	Banner ad.
20	30	10
30	80	30
50	40	50
100	70	40
50	150	90

5. A medical researcher is testing the effectiveness of various types of cold medicine for children. Specifically, he wants to find out whether the effects of one type of cold medicine lasts longer than the effects of other types of cold medicine. He selects a random sample of children with a cold from a local school and examines how long the effects of four types of cold medicine last.

A. Set up the research and null hypotheses.

B. Complete the following ANOVA table.

Source	Sum of Squares	Degrees of Freedom	Mean Sum of Squares	F
Between	400	?	?	?
Within	?	90	?	
Total	2,300	?		

C. What is the critical f value?

D. Do an F test with the information above.

E. Calculate and interpret an eta-squared.

6. A human resources manager of a large company is interested in finding out about average salaries of employees with varying education levels. He thinks employees with more education have higher salaries than employees with less education. He collects the following salaries from 9 employees in the Marketing Department. Answer the following questions.

Annual Salaries by Education

High School Diploma	Some College	Bachelors
$20000	$30000	$70000
$30000	$50000	$80000
$40000	$40000	$90000

A. What is the research hypothesis?

B. What is the null hypothesis?

C. What is the independent variable?

D. What is the dependent variable?

E. At what level is the independent variable measured?

F. At what level is the dependent variable measured?

G. Do an f-test.

7. A telecommunications analyst is interested in finding out who pay more for monthly cell phone bills. She is particularly interested in comparing the usage of three different types of cell phone users— those who use gaming applications, those who use social networking applications, and those who use neither of these two applications. She thinks that individuals who use gaming applications pay more than other individuals. By surveying a sample of 12 individuals in a region, she obtains the following information. Do an f-test.

Average amount of payments in dollars by cell phone users

Use gaming	Use social networking	Use neither
$ 60	$60	$20
$100	$60	$60
$ 80	$80	$40
$160	$80	$40

8. A medical researcher at a non-profit hospital thinks that the value of a healthy diet and the power of prayer should be taken into consideration when doctors discuss treatment options with their patients. To determine whether the healthy diet and the power of prayer help increase cancer remission rates, he designs an experiment that divides 100 randomly selected cancer patients into three equal size groups: those who receive regular treatments, those who receive regular treatments with a special diet program, and those who receive regular treatments with a prayer program. The diet program puts patients on a low calorie, high fiber, mostly vegetable diet; and the prayer program encourages patients to pray and their doctors frequently emphasizes the power of prayer. He expects that the remission rates of the patients who receive regular treatments only are lower than the other two groups. Set up the research and null hypotheses. Complete the following ANOVA table, and do an f-test.

Source	Sum of Squares	Degrees of Freedom	Mean Sum of Squares	F
Between	1000	?	?	?
Within	?	?	?	
Total	3,000	?		

9. A technology company manager is exploring the possibility of manufacturing smart phones in another country. She informally surveys 9 companies in three different countries and finds the following information. Do an f-test.

Country A	Country B	Country C
100	90	90
120	80	110
110	100	100

10. A telecommunications analyst is interested in finding out whether a new start up prepaid cell phone service company's customers pay less than those of its competitors. She is interested in the actual amounts of cell phone bills rather than the amounts that each company claims that it charges. She thinks that the new company's customers are paying less than those of its competitors. She picks two other prepaid cell phone service companies and surveys 9 actual customers of the three companies. Do an f-test.

Monthly Bills Paid by Customers of Three Prepaid Companies

New Company	Company X	Company Y
$30	$50	$60
$40	$40	$40
$20	$60	$50

Chapter 10

Crosstabulation Analysis

1. A visionary leader provides a clear picture of a desirable future of an organization. He or she leads the organization in getting to the desirable future with imagination, foresight, and creativity. As a result, members of the organization are excited about the future of the organization and are willing to commit to the success of the organization. As a management analyst, you think businesses led by a more visionary leader are less likely to fail than businesses led by a less visionary leader. The following is a table with the degree of vision and business failure. Calculate column percentages and interpret a column percentage difference.

	Degree of Vision	
Business Failure Status	Less visionary	More visionary
Failure	50	40
Success	50	60

2. According to the U.S. Census Bureau (2012), there are about 29.4 million business establishments in the United States and close to 48,000 companies filed for bankruptcy protection in 2011, according to the American Bankruptcy Institute. You learn that plumbing, single family housing construction, grocery stores, independent restaurants, security brokers, local trucking companies are most likely to fail in and religious organizations, apartment buildings, vegetable crop productions, offices of medical doctors, child day care services are least likely to fail in five years, according to the Statistic Brain website. You are thinking about opening a restaurant and collects the following information from a local area. The business failure status variable refers to failure or success after 5 years of operation.

Calculate column percentages and interpret a column percentage difference.

	Type of restaurant	
	Independent restaurant	Chain restaurant
Business Failure Status		
Failure	45	30
Success	50	65

3. A business analyst considers the effect of competitive advantage on profit margins. She thinks that companies whose employees pursue competitive advantage more over rival companies using such strategies as cost leadership, differentiation, innovation, operational efficiency, and customer satisfaction have higher profit margins than companies whose employees pursue competitive advantage less over rival companies. A survey of 150 companies reveals that 60 of the 90 companies that pursue competitive advantage more actively have high profit margins and 30 of the 60 companies that pursue competitive advantage less actively have high profit margins. Construct a table between the degree of pursuit in competitive advantage and the level of profit margins, calculate column percentages, and interpret a column percentage difference.

4. A management analyst thinks that employees who report to a transformational leader are more satisfied with their jobs than employees who report to a transactional leader. Transformational leaders inspire their followers by generating admiration from their followers, motivating them through inspiration, intellectually challenging them to be innovative, and addressing their needs by paying personal attention to them. Transactional leaders motivate followers using rewards and punishments, rewarding those who perform well and punishing those who perform poorly. A survey of 80 employees shows that 30 of the 40 employees who report to a transformational leader are highly satisfied with their jobs and 22 of the 40 employees who report to a transactional leader are highly satisfied with their jobs. Construct a table between leadership style and job satisfaction levels, calculate column percentages, and interpret a column percentage difference.

5. Leaders who practice the leader-member exchange theory divide, either intentionally or unintentionally, employees into the in-group that receive more attention, information, and privileges and the out-group that receive less attention, information, and privileges. As a management analyst, you think that in–group employees are more likely to feel a sense of belonging to their company than out-group employees. A survey of 400 employees reveal that 90 of the 100 in-group employees feel a strong sense of belonging to the company and 150 of the 300 out-group employees feel a strong sense of belonging to the company. Construct a table between employees' in-group status and sense of belonging, calculate column percentages, and interpret a column percentage difference.

6. A manager thinks that employees who interact with their managers more often trust the management of their company more than employees who interact with their managers less often. His survey of 300 employees reveals that 110 of the 150 employees who interact with their managers more often highly trust the management of their company and 90 of the 150 employees who interact with their managers less often highly trust the management of their company. Construct a table between employees' interaction status and sense of belonging, calculate column percentages, and interpret a column percentage difference.

7. A management analyst thinks that employees who have more job autonomy are more satisfied with their jobs than employees who have less job autonomy. For the purpose of his study, he defines job autonomy as the ability to perform a job without much supervision. They include the discretion and freedom to determine job responsibilities, perform tasks, and schedule work hours. In order to test his hypothesis, he surveys 100 employees. His survey reveals that

45 of the 50 employees with a high level of job autonomy are highly satisfied with their jobs and 35 of the 50 employees with a low level of job autonomy are highly satisfied with their jobs. Construct a table between job autonomy and job satisfaction, calculate column percentages, and interpret a column percentage difference.

8. A management analyst thinks that companies that use a Strengths, Weaknesses, Opportunities, and Threats (SWOT) analysis explicitly have higher profit margins than companies that do not use a SWOT analysis explicitly. In order to test her hypothesis, she surveys 120 companies in one industry and measures SWOT analysis status and levels of profit margins. Her survey reveals that 40 of the 50 companies that use a SWOT analysis explicitly have high profit margins and 45 of the 70 companies that do not use a SWOT analysis explicitly have high profit margins. Construct a table, calculate column percentages, and interpret a column percentage difference.

9. A manager is wondering about the effects of Porter's generic strategies on profit margins. She thinks that the differentiation strategy (providing different products from competitors) is more effective than the cost leadership strategy (providing products at lower costs than competitors) at the high end segment of the market. Her research indicates that 70 of the 120 companies that use a differentiation strategy have high profit margins and 20 of the 70 companies that use a cost leadership strategy have high profit margins. Construct a table, calculate column percentages, and interpret a column percentage difference.

10. A marketing manager thinks that companies that sell more unique products that are difficult to imitate generate more sales than companies that sell less unique products. She examines various sales data of 200 similar size companies in one industry and finds that 50 of the 80 companies that sell more unique products have a high amount of sales and 50 of the 120 companies that sell less unique products have a high amount of sales. Construct a table, calculate column percentages, and interpret a column percentage difference.

Chapter 11

Crosstabulation Analysis with a Chi Square Test

1. For Question 1 in Chapter 10, do a chi square test.

2. For Question 2 in Chapter 10, do a chi square test.

3. For Question 3 in Chapter 10, do a chi square test.

4. For Question 4 in Chapter 10, do a chi square test.

5. For Question 5 in Chapter 10, do a chi square test.

6. For Question 6 in Chapter 10, do a chi square test.

7. For Question 7 in Chapter 10, do a chi square test.

8. For Question 8 in Chapter 10, do a chi square test.

9. For Question 9 in Chapter 10, do a chi square test.

10. For Question 10 in Chapter 10, do a chi square test.

Chapter 12

Measures of Association

1. Go to Question 1 in Chapter 10. Calculate and interpret an appropriate measure of association for Question 1.

2. Go to Question 2 in Chapter 10. Calculate and interpret an appropriate measure of association for Question 2.

3. Go to Question 3 in Chapter 10. Calculate and interpret an appropriate measure of association for Question 3.

4. Go to Question 4 in Chapter 10. Calculate and interpret an appropriate measure of association for Question 4.

5. Go to Question 5 in Chapter 10. Calculate and interpret an appropriate measure of association for Question 5.

6. Go to Question 6 in Chapter 10. Calculate and interpret an appropriate measure of association for Question 6.

7. Go to Question 7 in Chapter 10. Calculate and interpret an appropriate measure of association for Question 7.

8. Go to Question 8 in Chapter 10. Calculate and interpret an appropriate measure of association for Question 8.

9. Go to Question 9 in Chapter 10. Calculate and interpret an appropriate measure of association for Question 9.

10. Go to Question 10 in Chapter 10. Calculate and interpret an appropriate measure of association for Question 10.

Chapter 13

Simple Regression

1. James sells printers on eBay. He wants to assess the effects of the number of printers sold per week on his weekly net profit. After collecting weekly data for a year, he constructs the following regression equation.

Amount of net profit (in dollars) = −30 + 15 Number of printers sold
N = 52 R-square = .90.

A. What is the independent variable in the equation?

B. What is the dependent variable in the equation?

C. What is the value of the regression coefficient?

D. What is the value of the y-intercept?

E. Interpret the regression coefficient.

F. Interpret the y-intercept.

G. Interpret the r-square.

H. When the number of printers sold is 10, what is the predicted amount of net profit?

2. Jasmine is interested in assessing the effects of the number of smart phones sold on the amount of net profits her company receives from the smart phone sales. After collecting weekly data for a year, she constructs the following regression equation.

Amount of net profit (in dollars) = -50000 + 100 Number of smart phones sold
N = 52 R-square = .80

A. Interpret the regression coefficient, y-intercept, and R-square.

B. What is the predicted amount of net profit when the number of smart phones sold is 1000?

3. A marketing manager of a computer tablet manufacturing company is interested in assessing the effects of advertising on the sales of tablets. She postulates that the amount of advertisement expenditures is positively related to the number of tablet sales. She randomly selects 50 regional advertising markets and obtains the amount of ad expenditures (in dollars) and the number of tablet sales in those markets. She then runs the data in SPSS and reports the following information.

Number of tablet sales = 50 + .2 Amount of ad expenditures
N = 50. R-square = .55

A. Interpret the regression coefficient, y-intercept, and R-square.

B. What is the predicted number of tablet sales when the amount of advertisement expenditures is 1000 dollars?

4. A marketing manager of a computer tablet manufacturing company is interested in assessing the effects of advertising on the amount of tablet sales in dollars. She postulates that the amount of advertisement expenditures is positively related to the amount of tablet sales. She randomly selects 50 regional advertising markets and obtains the amount of ad expenditures (in dollars) and the amount of tablet sales

in dollars in those markets. She then runs the data in SPSS and reports the following information.

Amount of tablet sales in dollars $= 20000 + 90$ Amount of ad expenditures
$N = 50$. R-square $= .72$

A. Interpret the regression coefficient, y-intercept, and R-square.

B. What is the predicted amount of tablet sales in dollars when the amount of advertisement expenditures is 1000 dollars?

5. An insurance company manager thinks that the number of vehicle insurance claims filed by drivers is negatively related to the company's net profits. She collects the information on the quarterly claims data and the amount of quarterly net profit (in dollars) in the last two years and regresses the latter on the former.

Amount of net profit $= 1500000 - 7,000$ Number of claims processed
R-square $=.61$ $N = 24$

A. Interpret the regression coefficient, y-intercept, and R-square.

B. What is the predicted amount of net profit when the number of vehicle insurance claims is 100?

6. Many financial companies including Morningstar publish fair value estimates of major companies. Fair value is an estimate of a company's value per share that is calculated from its future growth rates, profit margins, risk factors, and other items. A stock analyst reports the following regression equation between fair value estimates in dollars and share prices in dollars.

Share price $= 1 + 1.2$ Fair value
R-square $=.95$ $N = 100$

A. Interpret the regression coefficient, y-intercept, and R-square.

B. What is the predicted price of a company's stock when the fair value is estimated to be $100?

7. The quality control director of a semi-conductor manufacturing company wants to know whether employees' job experience is related to the number of defective wafers produced (perhaps because lack of job experience leads to more defective wafers). She collects data for a year and regresses the number of defective wafers from the individual employees on their job experience (number of years at job). Assume that the regression coefficients are statistically significant at the .05 level and the company produces only one kind of wafer. Interpret the regression coefficient, Y-intercept, and R-square.

Number of Defective Wafers = 100 − 4 Job Experience
N = 30   R-square = .67

8. State a regression equation with two variables (one independent variable and one dependent variable) that you deal with in your workplace. The regression equation should have a y-intercept, regression coefficient, R-square, and N (number of cases). Interpret the y-intercept, regression coefficient, and R-square in the regression equation.

9. Regression analysis is often used to estimate the cost of production with some cost factors. The following information is obtained from a cell phone company's production Department.

Number of smart phones produced (units)	Total cost of production ($)
1	$20
3	$55
4	$70
6	$100
10	$150

A. Develop a scatter gram with production units as the independent variable and production cost as the dependent variable.

B. Develop a regression equation that could be used to predict the amount of total production cost based on the number of units produced. More specifically, calculate the regression coefficient, y-intercept, and R-square.

C Interpret the regression coefficient, y-intercept, and R-square.

D. What is the predicted amount of production cost when the number of smart phones produced is 5?

10. State real or hypothetical data with two variables — one independent and one dependent. An example is in Question 9. Then answer the following questions.

A. Develop a scatter gram with the independent variable and the dependent variable.

B. Develop a regression equation that could be used to predict the dependent variable based on the independent variable. More specifically, calculate the regression coefficient, y-intercept, and R-square.

C Interpret the regression coefficient, y-intercept, and R-square.

D. Predict the value of the dependent for a given value of the independent variable.

II. Suggested Answers to Odd-Numbered Questions

Chapter 1

Hypothesis

1.

A. Employees with a high level of emotional intelligence do not have fewer conflicts with other employees in the workplace than are individuals with a low level of emotional intelligence.

B. Employees.

C. Levels of emotional intelligence.

D. Number of conflicts employees have with other employees in the work place.

3.

A. Hospitals headed by a transformational leader provide a more satisfactory work environment than hospitals headed by a transactional leader.

B. Hospitals headed by a transformational leader do not provide a more satisfactory work environment than hospitals headed by a transactional leader.

C. Leadership style (transformational, transactional).

D. Levels of satisfactory work environment.

5.

Answers will vary. The following is an example.

Research hypothesis: Graduates with good quantitative skills get higher paying jobs than graduates with poor quantitative skills.

Null hypothesis: Graduates with good quantitative skills do not get higher paying jobs than graduates with poor quantitative skills.

Cases: Graduates.

Independent variable: Degree of quantitative skills.

Dependent variable: levels of pay.

7.

Research hypothesis: Home loan applicants with high credit scores are more likely to exercise a strategic default than home loan applicants with low credit scores.

Null hypothesis: Home loan applicants with high credit scores are not more likely to exercise a strategic default than home loan applicants with low credit scores.

Cases: Home loan applicants.

Independent variable: Levels of credit score

Dependent variable: Likelihood to exercise a strategic default.

9.

Research hypothesis: Customers who use a coupon at the store purchase more items than customers who do not use a coupon at the store.

Null hypothesis: Customers who use a coupon at the store do not purchase more items than customers who do not use a coupon at the store.

Cases: Customers.

Independent variable: Coupon usage (yes, no).

Dependent variable: Number of items purchased.

Chapter 2

Levels of Measurement

1.

A. Interval-ratio (categories, ranking, and equal intervals).

B. Nominal (categories only and no ranking).

C. Nominal (categories only and no ranking).

D. Ordinal (categories and ranking, but no equal intervals).

E. Nominal (categories only and no ranking).

F. Interval-ratio (categories, ranking, and equal intervals).

G. Ordinal (categories and ranking, but equal intervals).

H. Ordinal (categories and ranking, but no equal intervals).

3.

A. Levels of self-esteem.

B. Number of conflicts employees have with coworkers.

C. Ordinal (categories, ranking, but no equal intervals).

D. Interval-ratio (categories, ranking, and equal intervals).

5.

A. Companies led by a more visionary CEO overcome recessions better than companies with a less visionary CEO.

B. Levels of vision of CEOs.

C. Ability to overcome recessions.

D. Ordinal (categories, ranking, but no equal intervals).

E. Ordinal (categories, ranking, but no equal intervals).

7.

Independent variable: Usage of cloud computing (yes, no)

Dependent variable: Levels of efficiency in information technology operations.

Level of measurement of the Independent variable: Nominal (categories, but no ranking).

Level of measurement of the Dependent variable: Ordinal (categories, ranking, but no equal intervals).

9. Consider the following research hypothesis: Employees respect managers who privately criticize, but publically praise them more than managers who privately praise, but publically criticize them. State the independent and dependent variables and explain the level at which you would measure the independent and dependent variables.

Independent variable: Methods of communication (privately criticizing and publically praising vs. privately praising and publically criticizing).

Dependent variable: Levels of respect.

Level of measurement of the independent variable: Nominal (categories, but no ranking).

Level of measurement of the dependent variable: Ordinal (categories, ranking, but no equal intervals).

Chapter 3

Frequencies and Graphs

1. Would you draw a bar chart or a histogram for the following variable? Why?

A. Histogram because it is an interval-ratio variable.

B. Bar chart because it is an ordinal variable.

C. Histogram because it is an interval-ratio variable.

3.

Frequency Distribution of the Number of Toys Sold by
Different Toy Types

Toy Type	Frequency	Proportion	Cumulative proportion
Stardoll by Barbie	50	(.20)	(.20)
Furby	45	(.18)	(.38)
Spider-Man Shooting Figure	35	(.14)	(.52)
Twister Dance	25	(.10)	(.62)
All Others	100	(.39)	(1.01)*

*It is greater than 1.0 because of rounding errors.

5.

A. Histogram because it is an interval-ratio variable.

B. Bar chart because it is a nominal variable.

C. Histogram because it is an interval-ratio variable. .

7.
A. Symmetric.

B. Positively skewed.

C. Negatively skewed.

D. Single value.

E. Bimodal.

F. Uniform.

9.
1. Actuaries
Future job openings as a percentage of 2010 employment: 87.1%
New Openings, 2010 to 2020: 18,900

2. Glaziers (glass installers)
Future job openings as a pct. of 2010 employment: 79.7%
New openings, 2010 to 2020: 33,400

3. Statisticians
Future job openings as a pct. of 2010 employment: 74.5%
New openings, 2010 to 2020: 18,700

4. Pest Control Workers
Future job openings as a pct. of 2010 employment: 70.9%
New openings, 2010 to 2020: 48,500

5. Interpreters and Translators
Future job openings as a pct. of 2010 employment: 69%
New openings, 2010 to 2020: 40,300

6. Optometrists
Future job openings as a pct. of 2010 employment: 68.4%

New openings, 2010 to 2020: 23,400

7. Natural Science Managers
Future job openings as a pct. of 2010 employment: 68%
New openings, 2010 to 2020: 33,500

8. Market Research Analysts and Marketing Specialists
Future job openings as a pct. of 2010 employment: 67.8%
New openings, 2010 to 2020: 191,800

9. Insulation Workers
Future job openings as a pct. of 2010 employment: 67.5%
New openings, 2010 to 2020: 34,700

10. Environmental Science and Protection Technicians, Including Health
Future job openings as a pct. of 2010 employment: 65.9%
New openings, 2010 to 2020: 19,500

Frequency Distribution of Top Ten New Job Openings

Job Opportunities	Frequency	Proportion	Cumulative Proportion
Actuaries	18,900	(.04)	(.04)
Glaziers	33,400	(.07)	(.11)
Statisticians	18,700	(.04)	(.15)
Pest Control Workers	48,500	(.11)	(.26)
Interpreters	40,300	(.09)	(.35)
Optometrists	23,400	(.05)	(.40)
Natural Science Managers	33,500	(.07)	(.47)
Marketing Analysts	191,800	(.42)	(.89)
Insulation Workers	34,700	(.08)	(.97)
Environmental Technicians	19,500	(.04)	(1.01)*

* It is greater than 1.0 because of rounding errors.

Chapter 4

Central Tendency and Dispersion Measures

1. A. 7   B. 4.

3. The customer service ratings of the SweepingFeet unit were typically 0 away from the mean of 80 (all the ratings were 80). The customer service ratings of the GoingExtra Miles unit were typically 20 away from the mean of 80. In the SweepingFeet unit, everyone received a rating of 80. In the GoingExtra Miles unit, some received ratings that were well above the mean of 80, whereas others received ratings that were well below the mean of 80. The customer service ratings of the SweepingFeet unit were completely homogeneous and the customer service ratings of the GoingExtra Miles unit were more heterogeneous.

5.

A. The salary distribution of the White Gold corporation is symmetric, whereas the salary distribution of the Yellow Gold corporation is positively skewed.

B. The salaries of the employees of the White Gold corporation are typically $10,00 away from the mean of $80,000. The salaries of the employees of the Yellow Gold corporation are typically $70,00 away from the mean of $80,000.

C. In the White Gold corporation, most employees make close to $80,000. In the Yellow Gold corporation, some employees make a lot more than $80,000 and other employees make a lot less than $80,000. The salaries of the employees of the White Gold corporation are more homogeneous and the salaries of the employees of the Yellow Gold corporation are more heterogeneous.

7.

A. The median is 770. $8+1 = 9/2 = 4.5$. The median is between the 4th observation and the 5th observation. $777 + 763 = 1540/2 = 770$.

B. 768.75 ($6150/8 = 768.75$).

C. 39.57. The customer satisfaction ratings of the eight company's devices are typically 39.57 points away from the mean of 768.75 points.

$(849-768.75)^2 = (80.25)^2 = 6440.06$
$(790-768.75)^2 = (21.25)^2 = 451.56$
$(782-768.75)^2 = (13.25)^2 = 175.56$
$(777-768.75)^2 = (8.25)^2 = 68.06$
$(763-768.75)^2 = (-5.75)^2 = 33.06$
$(742-768.75)^2 = (-26.75)^2 = 715.56$
$(740-768.75)^2 = (-28.75)^2 = 826.56$
$(707-768.75)^2 = (-61.75)^2 = 3813.06$
$12523.48/8 = 1565.44$. $\sqrt{1565.44} = 39.57$.

9.

A. Find the median of the dividend yield. 4.10. $10+1 = 11/2 = 5.5$. The median is between the 5th observation and the 6th observation. $4.11+ 4.09 = 8.2/2 = 4.1$.

B. Calculate the mean.4.28. $42.77/10 = 4.28$.

C. Calculate and interpret the standard deviation. The standard deviation is .53. The dividends are typically .53% away from the mean of 4.28%.

$5.39 - 4.28 = (1.11)^2 = 1.23$

$4.98 - 4.28 = (.7)^2 = .49$
$4.64 - 4.28 = (.36)^2 = .13$
$4.24 - 4.28 = (-.04)^2 = .00$
$4.11 - 4.28 = (-.17)^2 = .03$
$4.09 - 4.28 = (-.19)^2 = .04$
$4.06 - 4.28 = (-.22)^2 = .05$
$3.81 - 4.28 = (-.47)^2 = .22$
$3.79 - 4.28 = (-.49)^2 = .24$
$3.66 - 4.28 = (-.62)^2 = .38$

$2.81/10 = .28. \sqrt{.28} = .53.$

Chapter 5

The Normal Distribution

1. .0228. $Z = (500-300)/100 = 200/100 = 2$. The probability = .0228.

3. .0014. $Z = (11-5) = 6/2 = 3$. The probability = .0014.

5.
A. .0228. $Z = (1600-1000) = 600/300 = 2$. The probability = .0228.

B. .0014. $Z = (100-1000) = -900/300 = -3$. The probability = .0014.

C. .14%. $Z = (1900-1000) = 900/300 = 3$. The probability = .0014. The percentage is .14. It is obtained by: $.0014 \times 100 = .14$.

D. 228. $Z = (400-1000) = -600/300 = 2$. The probability = .0228. The number of cases is 228. It is obtained by: $.0228 \times 10000 = 228$.

7.
A. .0014. $Z = (820-700)\ 120/40 = 3$. The probability = .0014.

B. About 1 ($.0014 \times 1000 = 1.4$).

C. .0228. $Z = (620-700) = -80/40 = -2$. The probability = .0228.

D. About 23 ($.0228 \times 1000 = 22.8$).

E. Yes. Her score is in the top 2.28%, which is within the top 10%. $Z = (780-700) = 80/2 = 2$. The probability = .0228. The percentage is 2.28%.

9. The following are the average prices of small (7 inch) tablet computers. Assume that the prices are normally distributed.

A. 171. $1710/10 = 171$.

B. The standard deviation is 73. The average tablet prices are typically 73 dollars away from the mean of 171 dollars.

$200 - 171 = (29)^2 = 841$
$320 - 171 = (149)^2 = 22201$
$200 - 171 = (29)^2 = 841$
$200 - 171 = (29)^2 = 841$
$200 - 171 = (29)^2 = 841$
$180 - 171 = (9)^2 = 81$
$100 - 171 = (71)^2 = 5041$
$180 - 171 = (9)^2 = 81$
$70 - 171 = (101)^2 = 10201$
$60 - 171 = (111)^2 = 12321$

$53290/10 = 5329$. $\sqrt{5329} = 73$.

C. .40. The average price of $200 for Samsung Galaxy Tab is .40 standard deviations away from the mean of $171. $Z = (200 - 171)/73 = 29/73 = .40$.

D. -.97. The average price of $100 for Polaroid Tablet is .97 deviations below the mean of $171. $Z = 100 - 171 = -71/73 = -.97$

E. .0014. What is the probability of finding a tablet that is more than $390? $Z = 390 - 171 = 219/73 = 3$. The probability is .0014.

F. .0228. What is the probability of finding a tablet that is less than $25 in this distribution?
$Z = 25 - 171 = -146/73 = -2$. The probability = .0228.

Chapter 6

Optional Chapter: Sampling, Sampling Distribution, and Central Limit Theorem

1. 71,000 ± 196. We are 95% confident that this interval captures the true population mean.

3. 12 ± .39. We are 95% confident that this interval captures the true population mean

5. 50000 ± 2772.28. We are 95% confident that this interval captures the true population mean.

7. 900 ± 6.94. We are 95% confident that this interval captures the true population mean.

9. 199 ± 5.06. We are 95% confident that this interval captures the true population mean. $208 is outside the 95% confidence interval. Therefore, she should adjust the price.

Chapter 7

Population Mean Compatibility Test: Hypothesis Testing with One Sample Mean

1.

1. The research hypothesis is that the average salary of employees is lower than the average salary of all engineering company employees in the state ($\mu < \$70,000$).

2. The null hypothesis is that the average salary of employees is not lower than the average salary of all engineering company employees in the state ($\mu \geq \$70,000$).

3. This is a one-tail test situation because we expect the sample mean to be lower than (not just different from) the hypothesized population mean.

4. We can do a z-test because the sample size is at least 100.

5. The standard error formula is **s/√n**. The standard error is: $(7000/\sqrt{100}) = 7000/10 = 700$.

6. The test statistic formula is: **(Sample mean - μ)/Standard error**. The test statistic (z score in this test) is -7.14. It is obtained by: $(65,000 -70000)/700 = -7.14$.

7. The rejection z score is -1.65 and the rejection region is the area beyond the z score of -1.65.

8. Since the test statistic, −7.14, falls in the rejection region (it is lower than −1.65), we reject the null hypothesis and claim the research hypothesis.

9. We claim that the average salary of employees is lower than the industry average in the state. When we claim this, we would be wrong less than 5% of the time.

3.
1. The research hypothesis is that the average number of copies sold per store this month is higher than that of last month ($\mu > 300$).

2. The null hypothesis is that the average number of copies sold per store is not higher this month ($\mu \leq 300$).

3. We need to do a one-tailed test because we are expecting that the average this month is higher than (not just different from) that of last month.

4. We need to do a t-test because the sample size is lower than 100. The sample size is 50.

5. $(30/\sqrt{50}) = 30/7.07 = 4.24$.

6. The value of the test statistic is $[(307 - 300 = 7) / 4.24] = 1.65$.

7. The degree of freedom is 49 (50 − 1). The rejection t value with an alpha of .05 and DF = 40 is 1.684. So the rejection region is the area beyond the rejection t value of 1.684.

8. We fail to reject the null hypothesis because the test statistic of 1.65 does not fall in the rejection region.

9. We cannot claim that the average number of copies sold this month is higher than last month's. Perhaps the album should be marketed more effectively.

5.

1. The research hypothesis is that senior citizens' average recovery time from the common cold is different from the average recovery time of the city's population as a whole (H_R: $\mu \neq 10.00$).

2. The null hypothesis is that senior citizens' average recovery time from the common cold is NOT different from the average recovery time of all adult city residents. (H_0 : $\mu = 10.00$).

3. We will do a two-tailed test and use both sides of the normal distribution.

4. Because our sample size is 100, we do a z test.

5. The standard error is obtained by **[s/√n]**. It is: ($2/\sqrt{100}$) = .2.

6. 5. It is obtained by: **(Sample mean - μ)/Standard error.** (11-10)/.2 = 1/.2 = 5.

7. The rejection region is the area beyond a z score of -1.96 and the area beyond a z score of +1.96.

8. Because the test statistic of 5 falls in the rejection region, we reject the null hypothesis of no difference and say that the sample mean of 11 is significantly different from the population mean of 10.00.

9. We conclude that senior citizens' average recovery time is significantly different from the average recovery time of the city's population as a whole. When we conclude this we would be wrong less than 5% of the time (or we are 95 percent confident that the sample mean is different from the population mean).

7.

1. The company takes more than 60 days to process invoices.

2. The company does not take more than 60 days to process invoices.

3. We will do a one-tail test because we expect the company to take more than 60 days to process invoices.

4. Because our sample size is at least 100, we do a z test.

5. .82. The standard error formula is: **[s/√n]**. $10/\sqrt{150} = 10/12.25 = .82$.

6. 12.20. The test statist formula is: **(Sample mean - μ)/Standard error.** $(70-60)/.82 = 12.20$.

7. The rejection region is the area beyond a z score of -1.65.

8. We reject the null hypothesis and claim that the sample mean of 70 is significantly higher than the population mean of 60.00.

9. We conclude that the company takes more than 60 days to process invoices. When we conclude this we would be wrong less than 5% of the time.

9.

1. The average monthly premium for the company in the pilot HMO program is lower than the average premium that the company paid for its employees last year.

2. The average monthly premium for the company in the pilot HMO program is not lower than the average premium that the company paid for its employees last year.

3. We will do a one-tail test because we expect that the monthly premium to be lower than last year's.

4. Because our sample size is lower than 100, we will do a t-test.

5. The standard error is obtained by [s/√n]. $50/\sqrt{40} = 7.91$. The standard error is 7.91.

6. The test statist formula is: **(Sample mean - μ)/Standard error.** $(490-500)/7.91 = -10/7.91 = -1.26$. The test statistic is -1.26.

7. The rejection region is the area beyond a t score of -1.697 with a df of 30 and an alpha of .05.

8. Because the test statistic of -1.26 does not fall in the rejection region, we fail to reject the null hypothesis.

9. We cannot claim that the average monthly premium for the company in the pilot HMO program is lower than the average premium that the company paid for its employees last year.

Chapter 8

Difference of Means Testing: Hypothesis Testing with Two Sample Means

1.

1. Costs of manufacturing smart phones by international companies are lower than those by U.S. companies.

2. Costs of manufacturing smart phones by international companies are not lower than those by U.S. companies.

3. Region (international, U.S.).

4. Costs of manufacturing smart phones.

5. Nominal.

6. Interval-ratio.

7. One-tail test because manufacturing costs by international companies are expected to be lower than (not just different from) those by U.S. companies.

8. Z-test because the sample size is at least 100.

9. The standard error formula is: $\sqrt{[s_1^2/n_1 + s_2^2/n_2]}$. $\sqrt{[40^2/52 + 50^2/50]}$ = $\sqrt{[1600/52 + 2500/50]}$ = $\sqrt{[30.77 + 50]}$. = $\sqrt{[80.77]}$ = 8.99.

10. The rejection region is the area beyond a z score of -1.65.

11. Test statistic: **(First Sample Mean – Second Sample Mean)/Standard Error**. $Z = (200 – 250)/ 8.99 = -50/ 8.99 = -5.56$

12. The null hypothesis is rejected because the test statistic of -5.56 falls in the rejection region.

13. Costs of manufacturing smart phones by international companies are lower than those by U.S. companies. When we claim this, we would be wrong less than 5% of the time.

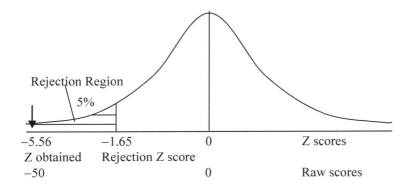

Rejection Region

5%

−5.56	−1.65	0	Z scores
Z obtained	Rejection Z score		
−50		0	Raw scores

3.

1. Stores that adopt a nostalgia marketing (NM) strategy have a higher amount of sales increase (in dollars) than stores that do not adopt a nostalgia marketing strategy.

2. Stores that adopt a nostalgia marketing (NM) strategy do not have a higher amount of sales increase (in dollars) than stores that do not adopt a nostalgia marketing strategy.

3. Status of adopting a nostalgia marketing (NM) strategy (yes, no).

4. An amount of sales increase.

5. Nominal.

6. Interval-ratio.

7. One-tail test because we expect that stores that adopt a nostalgia marketing (NM) strategy have a higher amount of sales increase than, not just a different amount of sales increase from, stores that do not adopt a nostalgia marketing strategy.

8. A z-test because the sample size is at least 100.

9. The standard error formula is: $\sqrt{[s_1^2/n_1 + s_2^2/n_2]}$. $\sqrt{[3000^2/60 + 1000^2/60]} = \sqrt{[9000000/60 + 1000000/60]} = \sqrt{[150000 + 16666.67]} = 408.25$.

10. The rejection region is the area beyond a z score of 1.65.

11. Test statistic: **(First Sample Mean – Second Sample Mean)/Standard Error.** $(10000-2000)/408.25 = 19.60$.

12. We reject the null hypothesis because the test statistic of 19.60 falls in the rejection region.

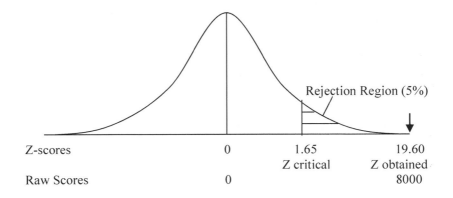

Z-scores	0	1.65	19.60
		Z critical	Z obtained
Raw Scores	0		8000

Rejection Region (5%)

13. We conclude that stores that adopt a nostalgia marketing (NM) strategy have a higher amount of sales increase (in dollars) than stores that do not adopt a nostalgia marketing strategy. When we claim this, we would be wrong less than 5% of the time.

5.

1. Individuals who use gaming applications on their service pay more than individuals who do not use gaming applications on their service.

2. Individuals who use gaming applications on their service do not pay more than individuals who do not use gaming applications on their service.

3. Status of gaming application use.

4. Amount of monthly payment.

5. Nominal.

6. Interval-ratio.

7. One-tail test because we expect that individuals who use gaming applications on their service pay more than (not just pay a different amount from) individuals who do not use gaming applications on their service

8. A z-test because the sample size is at least 100.

9. The standard error formula is: $\sqrt{[s_1^2/n_1 + s_2^2/n_2]}$. $\sqrt{[20^2/120 + 10^2/180]} = \sqrt{[400/120 + 100/180]} = \sqrt{[3.33 + .56]} = \sqrt{[3.89]} = 1.97$.

10. The rejection region is the area beyond a z score of 1.65.

11. Test statistic: **(First Sample Mean – Second Sample Mean)/Standard Error.** $(70 - 50)/1.97 = 20/1.97 = 10.15$.

12. We reject the null hypothesis because the test statistic of 10.15 falls in the rejection region.

13. We conclude that individuals who use gaming applications on their service pay more than individuals who do not use gaming applications on their service. When we claim this, we would be wrong less than 5% of the time.

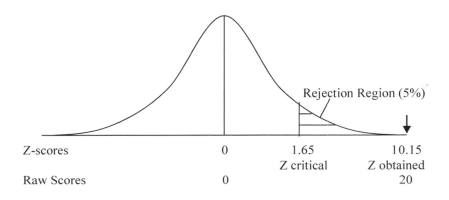

7.
1. Prepaid plan users pay less for unlimited talk, text, and data plans than postpaid plan users.

2. Prepaid plan users do not pay less for unlimited talk, text, and data plans than postpaid plan users.

3. Type of cell phone plan (pre-paid, post-paid).

4. An amount of monthly payment.

5. Nominal.

6. Interval-ratio.

7. One-tail test because we expect that prepaid plan users pay less for unlimited talk, text, and data plans than (not just pay a different amount from) postpaid plan users.

8. A z-test because the sample size is at least 100.

9. The standard error formula is: $\sqrt{[s_1^2/n_1 + s_2^2/n_2]}$. $\sqrt{[10^2/50 + 10^2/150]}$ = $\sqrt{[100/50 + 100/150]}$ = $\sqrt{[2 + .67]}$ = $\sqrt{[2.67]}$ = 1.63.

10. The rejection region is the area beyond a z score of -1.65.

11. Test statistic: **(First Sample Mean – Second Sample Mean)/Standard Error**. (50-100)/1.63 = -50/1.63 = -30.67.

12. We reject the null hypothesis because the test statistic of -30.37 falls in the rejection region.

13. We conclude that prepaid plan users pay less for unlimited talk, text, and data plans than postpaid plan users. When we claim this, we would be wrong less than 5% of the time.

9.
1. Stores that use Black Friday promotions have higher annual sales per square foot than stores that do not use Black Friday promotions.

2. Stores that use Black Friday promotions do not have higher annual sales per square foot than stores that do not use Black Friday promotions.

3. Black Friday promotion status (yes, no).

4. Amount of annual sales per square foot.

5. Nominal.

6. Interval-ratio.

7. A one-tail test because we expect that Stores that use Black Friday promotions have higher annual sales per square foot than (not just a different amount of sales from) stores that do not use Black Friday promotions.

8. A z-test because the sample size is at least 100.

9. The standard error formula is: $\sqrt{[s_1^2/n_1 + s_2^2/n_2]}$. $\sqrt{[80^2/60 + 90^2/40]}$ = $\sqrt{[6400/60 + 8100/40]}$ = $\sqrt{[106.67 + 202.50]}$ = $\sqrt{[309.17]}$ = 17.58.

10. The rejection region is the area beyond a z score of 1.65.

11. Test statistic: **(First Sample Mean – Second Sample Mean)/Standard Error**. (320 – 300)/ 17.58 = 20/17.58 = 1.14.

12. We fail to reject the null hypothesis because the test statistic of 1.14 does not fall in the rejection region.

13. We cannot conclude that stores that use Black Friday promotions have higher annual sales per square foot than stores that do not use Black Friday promotions.

Chapter 9

Analysis of Variance (ANOVA)

1.
A. At least one average hotel room price of the three cities is different from the average hotel room price of other cities.

B. The average hotel room prices of all three cities are the same.

C. Three cities.

D. Average hotel room price.

E. At the nominal level.

F. At the interval-ratio level.

G. We need the individual data points (average room prices of 20 hotels from each of the three cities, altogether 60 room prices).

3.
A. The job satisfaction score of at least one group of individuals is different from those of other groups of individuals.

B. The job satisfaction scores of all three groups of individuals are the same.

C. Three groups of individuals (those who work for themselves, those who work for someone else in large organizations, those who work for someone else in small organizations).

D. Job satisfaction scores.

E. At the nominal level.

F. At the interval-ratio level

G. We need the job satisfaction scores of 90 individuals, which will generate three means.

5.
A. The research hypothesis is that the effects of at least one medicine type last longer than the effects of other medicine types. The null hypothesis is that the effects of the four types of medicines last for the same length of time.

B.

Source	Sum of Squares	Degrees of Freedom	Mean Sum of Squares	F
Between	400	3	133.33	6.32
Within	1900	90	21.11	
Total	2,300	93		

The sum of squares within is 1900 (2,300 – 400). The degrees of freedom for the sum of squares between is 3 (4 – 1). The degrees of freedom for the total sum of squares is 93 (3 + 90). The mean sum of squares between is 133.33 (400/3). The mean sum of squares within is 21.11(1900/90). The f value is 6.32 (133.33/21.11).

C. The critical f value with an alpha of .05 and 3 and 60 (from 90) degrees of freedom is 2.76.

D. Because the f statistic of 6.32 is higher than the f critical value of 2.76, we reject the null hypothesis. We conclude that the effect of at least one medicine lasts longer than the effects of other medicines. When we say this, we would be wrong less than 5% of the time.

E. Eta-squared is .17 (400/2,300). Seventeen percent of the variation in the length of the effect of the medicine is explained by the medicine type.

7.

ANOVA Table

Source	Sum of Squares	Degrees of Freedom	Mean Sum of Squares	F
Between	7200	2	3600	4.76
Within	6800	9	755.56	
Total	14000	11		

Research hypothesis: The average monthly payment of at least one group of cell phone users is different from the average monthly payment of other groups. Null hypothesis: The average monthly payments of all three groups of cell phone users are the same.

The sum of square between is 7200, the sum of square within is 6800, the total sum of square is 14000.

The first mean is 100, the second mean is 70, and the third mean is 40. The grand mean is 70 (100 + 70 + 40 = 210/3 = 70). The total sum of squares is obtained by several steps. First we get the deviations between the individual score and the grand mean.

60-70 = -10   60 – 70 = -10   20 – 70 = -50
100-70 = 30   60 – 70 = -10   60 – 70 = -10
80 – 70 = 10   80 – 70 = 10   40 – 70 = -30
160 – 70 = 90 80 – 70 = 10   40 – 70 = -30

We square them and add them up. (100, 900, 100, 8100, 100, 100, 100, 100, 2500, 100, 900, 900) = 14000.

The sum of squares between is obtained by the following:
$100 - 70 = 30^2 = 900 \times 4 = 3600$
$70 - 70 = 0^2 \times 4 = 0$
$40 - 70 = (-30)^2 = 900 \times 4 = 3600$

$3600 + 0 + 3600 = 7200.$

The sum of squares within is: $14000 - 7200 = 6800$.

The degree of freedom for the total sum of squares is 11 (12-1 = 11), the degree of freedom for the sum of squares between is 2 (3-1 = 2), the degree of freedom for the sum of squares within is 9 (12-3 = 9).

The mean sum of squares between is 7200/2 = 3600. The mean sum of squares within is 6800/9 = 755.56. The f ratio is 3600/755.56 = 4.76.

The critical f score with degrees of freedom of 2 and 9 and with an alpha of .05 is 4.26. We reject the null hypothesis because the f ratio obtained of 4.76 is higher than the critical f score of 4.26. We conclude that the average monthly payment amount of at least one group of cell phone users is different from the average monthly payment amounts of other groups. When we claim this, we would be wrong less than 5% of the time.

9.
ANOVA Table

Source	Sum of Squares	Degrees of Freedom	Mean Sum of Squares	F
Between	600	2	300	3
Within	600	6	100	
Total	1200	8		

Research hypothesis: The average manufacturing cost of three companies in at least one country is different from that of other countries. Null hypothesis: The average manufacturing costs of three companies of all three countries are the same.

The sum of square between is 600, the sum of square within is 600, the total sum of square is 1200.

The first mean is 110, the second mean is 90, and the third mean is 100. The grand mean is 100.

The total sum of squares is obtained by the following process:
$100 – 100 = 0$ $90 -100 = -10$ $90 -100 = -10$
$120 – 100 = 20$ $80 – 100 = -20$ $110 -100 = 10$
$110 – 100 = 10$ $100 – 100 = 0$ $100 – 100 = 0$
$(0)^2 + (-10)^2 + (-10)^2 + (20)^2 + (-20)^2 + (10)^2 + (10)^2 + (0)^2 + (0)^2 = $ 1200.

The sum of squares between is obtained by the following process:
$110-100 = 10^2 = 100 \times 3 = 300$
$90 – 100 = (-10)^2 = 100 \times 3 = 300$
$100 – 100 = 0^2 = 0 \times 3 = 0$
$300 + 300 + 0 = 600.$

The degree of freedom for the total sum of squares is 8 (9-1 = 8), the degree of freedom for the sum of squares between is 2 (3-1 = 2), the degree of freedom for the sum of squares within is 6 (9-3 = 6).

The mean sum of squares between is: $600/2 = 300$. The mean sum of squares within is: $600/6 = 100$. The f ratio is 3.

The critical f score with degrees of freedom of 2 and 6 and with an alpha of .05 is 5.14. We cannot reject the null hypothesis because the f score obtained of 3 is not higher than the critical f score of 5.14. We cannot conclude that the average manufacturing cost of three companies in at least one country is different from that of companies in other countries.

Chapter 10

Crosstabulation Analysis

1.

1. Table construction. The table is already constructed.

	Degree of Vision	
Business Failure Status	Less visionary	More visionary
Failure	50	40
Success	50	60

2. Column percentages.

	Degree of Vision	
Business Failure Status	Less visionary	More visionary
Failure	50 50%	40 40%
Success	50 50%	60 60%

3. Column percentage difference interpretation.
Businesses led by a more visionary leader are 10% (50% - 40% = 10%) less likely to fail than businesses led by a less visionary leader.

3.

1. Table construction.

	Degree of Pursuit in Competitive Advantage	
Profit Margins	Less competitive advantage	More competitive advantage
Low	30	30
High	30	60

2. Column percentages.

Degree of Pursuit in Competitive Advantage

Profit Margins	Less competitive advantage	More competitive advantage
Low	30 50%	30 33%
High	30 50%	60 67%

3. Column percentage difference interpretation.
Companies that pursue competitive advantage more over rival companies are 17% (67% - 50% = 17%) more likely to have high profit margins than companies that pursue competitive advantage less over rival companies.

5.
1. Table construction.

	In-group employee status	
Sense of belonging	In-group	Out-group
Low	10	150
High	90	150

2. Column percentages.

	In-group employee status	
Sense of belonging	In-group	Out-group
Low	10 10%	150 50%
High	90 90%	150 50%

3. Column percentage difference interpretation.

In–group employees are 40% more likely to feel a sense of belonging to their company than out-group employees.

7.
1. Table construction.

	Job autonomy	
Job satisfaction	Low	High
Low	15	5
High	35	45

2. Column percentages.

	Job autonomy	
Job satisfaction	Low	High
Low	15 30%	5 10%
High	35 70%	45 90%

3. Column percentage difference interpretation.

Employees who have more job autonomy are 20% more satisfied with their jobs than employees who have less job autonomy.

9.

1. Table construction.

	Generic strategy type	
Profit margins	Cost leadership	Differentiation
Low	50	50
High	20	70

2. Column percentages.

Generic strategy type

Profit margins	Cost leadership	Differentiation
Low	50 71%	50 42%
High	20 29%	70 58%

3. Column percentage difference interpretation.

Companies that use the differentiation strategy are 29% more likely to have high profit margins than companies that use the cost leadership strategy.

Chapter 11

Crosstabulation Analysis with a Chi Square Test

1. Go to Question 1 in Chapter 10 and do a chi square test.
The chi square obtained is 2.02. It is lower than the rejection chi square of 3.841 (degree of freedom = 1, alpha = .05). Therefore, we fail to reject the null hypothesis. We cannot claim that businesses led by a more visionary leader are less likely to fail than businesses led by a less visionary leader.

3. For Question 3 in Chapter 10, do a chi square test.
The chi square obtained is 4.17. It is higher than the rejection chi square of 3.841 (degree of freedom = 1, alpha = .05). Therefore, we reject the null hypothesis and claim the research hypothesis. We conclude that companies that pursue competitive advantage more over rival companies are more likely to have high profit margins than companies that pursue competitive advantage less over rival companies. When we claim this, we would be wrong less than 5% of the time.

5. For Question 5 in Chapter 10, do a chi square test.
The chi square obtained is 50.00. It is higher than the rejection chi square of 3.841 (degree of freedom = 1, alpha = .05). Therefore, we reject the null hypothesis and claim the research hypothesis. We conclude that in–group employees are more likely to feel a sense of belonging to their company than out-group employees. When we claim this, we would be wrong less than 5% of the time.

7. For Question 7 in Chapter 10, do a chi square test.

The chi square obtained is 6.25. It is higher than the rejection   chi square of 3.841 (degree of freedom = 1, alpha = .05). Therefore, we reject the null hypothesis and claim the research hypothesis. We conclude that employees who have more job autonomy are more satisfied with their jobs than employees who have less job autonomy. When we claim this, we would be wrong less than 5% of the time.

9. For Question 9 in Chapter 10, do a chi square test.

The chi square obtained is 15.71. It is higher than the rejection   chi square of 3.841 (degree of freedom = 1, alpha = .05). Therefore, we reject the null hypothesis and claim the research hypothesis. We conclude that companies that use the differentiation strategy are more likely to have high profit margins than companies that use the cost leadership strategy. When we claim this, we would be wrong less than 5% of the time.

Chapter 12

Measures of Association

1. Go to Question 1 in Chapter 10. Calculate and interpret an appropriate measure of association for Question 1.

Lambda is 0. Knowledge of the vision status of the leader improves our ability to predict business failure by 0%. There appears to be no relationship between the two variables.

3. Go to Question 3 in Chapter 10. Calculate and interpret an appropriate measure of association for Question 3.

The gamma value is .33. There is a weak relationship between the pursuit of competitive advantage and profit margins. When we try to predict the order of pairs on profit margins, we would reduce our prediction error by 33% if we take the pursuit of competitive advantage into account.

5. Go to Question 5 in Chapter 10. Calculate and interpret an appropriate measure of association for Question 5.

Lambda is 0. Knowledge of the in-group status improves our ability to predict sense of belonging by 0%. There appears to be no relationship between the two variables.

7. Go to Question 7 in Chapter 10. Calculate and interpret an appropriate measure of association for Question 7.

The gamma value is .59. There is a moderate relationship between job autonomy and job satisfaction. When we try to predict the order of pairs on job satisfaction, we would reduce our prediction error by 59% if we take job autonomy into account.

9. Go to Question 9 in Chapter 10. Calculate and interpret an appropriate measure of association for Question 9.

Lambda is .22. Knowledge of the strategy type improves our ability to predict profit margins by 22%. There is a weak relationship between strategy types and profit margins.

Chapter 13

Simple Regression

1.
A. Number of printers sold.

B. Amount of net profit.

C. 15.

D. -30.

E. When the number of printers sold goes up by 1, the amount of profit goes up by 15$.

F. When the number of printers sold is 0, the amount of net profit is -30$.

G. The number of printers sold explains 90% of the variation in the amount of net profit.

H. $120.

3.
A. When the amount of ad expenditures goes up by 1 dollar, the number of tablet sales goes up by .2. When the amount of ad expenditures is 0, the number of tablet sales is 50. The amount of ad expenditures explains 55% of the variation in the number of tablet sales.

B. 250.

5.

A. When the number of claims processed goes up by 1, the amount of net profit goes down by $7,000. When the number of claims processed is 0, the amount of net profit is $15,000,000. The number of claims processed explains 61% of the variation in the amount of net profit.

B. $800,000.

7.

When job experience goes up by 1 year, the number of defective wafers goes down by 4. When job experience is 0, the number of defective wafers is 100. Job experience explains 67% of the variation in the number of defective wafers.

9.

A. Develop a scatter gram with production units as the independent variable and production cost as the dependent variable.

Scatter gram

Y: Total cost of production

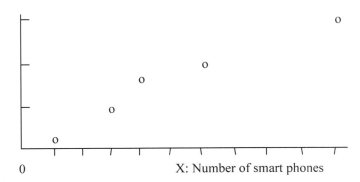

0   X: Number of smart phones

B. Develop a regression equation that could be used to predict the amount of total production cost based on the number of units produced. More specifically, calculate the regression coefficient, y-intercept, and R-square.

Regression coefficient = 14. 29
Y-intercept = 10.41 (This is obtained by the calculation process explained below. If we use a regression calculator on the web, we will get 10.38. The difference is due to rounding errors)
R square = .99

The regression equation is:
Total production cost = 10.41 + 14.29 Number of units
R square = .99

The regression coefficient is obtained by the following formula:

$$b = \frac{\Sigma\,(Xi - \text{Mean of X})\,(Yi - \text{Mean of Y})}{\Sigma\,(Xi - \text{Mean of X})^2}$$

Mean of X = 24/5 = 4.8 Mean of Y = 395/5 = 79.

$\Sigma\,(Xi - \text{Mean of X})^2$ is obtained by the following process:
1 − 4.8 = -3.8 $(-3.8)^2$ = 14.44
3 − 4.8 = -1.8 $(-1.8)^2$ = 3.24
4 − 4.8 = -.8 $(-.8)^2$ =   .64
6 − 4.8 = 1.2 $(1.2)^2$ = 1.44
10 − 4.8 = 5.2 $(5.2)^2$ = 27.04
The sum [$\Sigma\,(Xi - \text{Mean of X})^2$] is 46.8.

$\Sigma\,(Xi - \text{Mean of X})\,(Yi - \text{Mean of Y})$ is obtained by the following process:

(Xi – Mean of X)	(Yi – Mean of Y)
1 – 4.8 = -3.8	20 – 79 = -59
3 – 4.8 = -1.8	55 – 79 = -24
4 – 4.8 = -.8	70 – 79 = -9
6 – 4.8 = 1.2	100 – 79 = 21
10 – 4.8 = 5.2	150 – 79 = 71

(Xi – Mean of X) (Yi – Mean of Y)
224.2
43.2
7.2
25.2
369.2
The sum [Σ (Xi – Mean of X) (Yi – Mean of Y)] is 669.

b = 669/46.8 = 14.29. This is the regression coefficient.

The Y-intercept is obtained by the following formula:

Y intercept: a =   Mean of Y – b (Mean of X)
Y Intercept = 79 – 14.29 (4.8) = 79 – 68.59 = 10.41

The R square is obtained by the following formula:

R Square = Explained Variation/Total Variation

Total variation:
20 - 79 = $(-59)^2$ = 3481
55 - 79 = $(-24)^2$ =   576
70 – 79 = $(-9)^2$ =   81
100 – 79 = $(21)^2$ =   441
150 – 79 = $(71)^2$ = 5041
The sum is: 9620

Explained variation:

	Predicted Y – Mean of Y
When X = 1, Predicted Y = 24.7	-54.3
X = 3, Predicted Y = 53.28	-25.72
X = 4, Predicted Y = 67.57	-11.43
X = 6, Predicted Y = 96.15	17.15
X = 10, Predicted Y = 153.31	74.31

The sum of the squared deviations = $(-54.3)^2 + (-25.72)^2 + (-11.43)^2 + (17.15)^2 + (74.31)^2 = 2948.49 + 661.52 + 130.64 + 294.12 + 5521.98 = 9556.75$.

R Square = Explained Variation/Total Variation: 9556.75/9620 = .99.

C Interpret the regression coefficient, y-intercept, and R-square.
When the number of units produced goes up by 1, the total production cost increases by $14.29. When the number of units produced is 0, the total production cost is $10.41. The number of units produced explains 99% of the variation in the total production cost.

D. What is the predicted amount of production cost when the number of smart phones produced is 5? $81.86. It is obtained by: Total production cost = 10.41 + 14.29 (5) = 10.41 + 71.45 = $81.86.

III. Sample Exam Questions

Sample Mid-term Exam Questions

Question 1

Questions 1 - 4. Consider the following research hypothesis: Individuals with a high level of emotional intelligence are more successful in their professional lives than are individuals with a low level of emotional intelligence.

1. What is the null hypothesis?

Note: The null hypothesis is a statement that is the opposite of the research hypothesis. Usually, we add "not," "do not," or "does not" to the main verb of the sentence to turn a research hypothesis to its null hypothesis.

   ○   A.   Individuals

   ○   B.   Emotional Intelligence

   ○   C.   Individuals

   ⦿   D.   Individuals with a high level of emotional intelligence are not more successful in their professional lives than are individuals with a low level of emotional intelligence.

Question 2

2. What are the cases? Note: Cases are the observation units and they can be counted.

   ○   Success in professional life

○ Individuals with less emotional intelligence

⊙ Individuals

○ Individuals with more emotional intelligence

Question 3

3. What is the independent variable? Note: The independent variable is the cause and the dependent variable is the effect. The independent variable often comes before the dependent variable. At the same time, because the independent variable has more than one category, the second part (category) of the independent variable may come after the dependent variable.

○ Levels of success in professional life

○ Individuals

⊙ Levels of emotional intelligence

○ Types of intelligence (analytical, linguistic, ...)

Question 4

4. What is the dependent variable?

⊙ Levels of success in professional life

○ Types of emotional intelligence

○ Individuals

○ Status of emotion

Question 5

Questions 5 - 8. Determine the level of measurement (nominal, ordinal, or interval-ratio) of each of the following variables.

5. Status of participation in a leadership training program (participation, no participation)   Note: Nominal variables have categories only. Ordinal variables have categories and ranking. Interval-ratio variables have categories, ranking, and equal intervals.

- ○   Interval-ratio
- ○   Ordinal
- ⊙   Nominal
- ○   Richter Scale

Question 6

6. Levels of leadership experience (very experienced, somewhat experienced, neither experienced nor inexperienced, somewhat inexperienced, very inexperienced)

- ○   Nominal
- ⊙   Ordinal
- ○   Interval-ratio
- ○   Earth Mover Scale

Question 7

7. Level of financial management ability of managers (very capable, somewhat capable, neither capable nor incapable, somewhat incapable, very incapable)

- ○   Nominal
- ⊙   Ordinal
- ○   Interval-ratio
- ○   Ocean view scale

Question 8

8. Number of jobs individuals had

○ Nominal

○ Ordinal

◉ Interval-ratio

○ Mode

Question 9

Questions 9 - 12. Consider the following research hypothesis: Employees with a high level of self-esteem feel less frustration in the work place than individuals with a low level of self-esteem. She measures self-esteem levels with a 5 point scale question that has very low to very high categories and frustration levels with a 7 point scale question that has very low to very high categories.

9. What is the independent variable?

○ Employees

◉ Levels of self-esteem

○ Workplace

○ Seven point scale

Question 10

10. What is the dependent variable?

○ Organizations

◉ Levels of frustration in the work place

○ Levels of self-confidence

○ More self-esteem

Question 11

11. At what level is the independent variable measured?

○   Interval-ratio

◉   Ordinal

○   Nominal

○   Scale

Question 12

12. At what level is the dependent variable measured?

○   Nominal

◉   Ordinal

○   Interval-ratio

○   Interval

Question 13

Questions 13 - 16. The following are the means and standard deviations of the amounts of salaries of the employees of two corporations: Sky and Earth.

	Sky Corporation	Earth Corporation
Mean	$70,000	$70,000
Standard Deviation	$5,000	$50,000
N	100	100

13. Interpret the standard deviation of $5,000. Note: The format of the sentence that interprets the standard deviation is: (Name of the variable) is typically (amount of standard deviation) away from the mean of xxx. Here the variable name is the amounts of salaries, the standard deviation amount is $5,000, and the mean is $70,000.

◉   The standard deviation of $5,000 at the Sky Corporation indicates that the salaries of the employees are typically $5,000 away from the mean of $70,000 (from roughly $65,000 to $75,000).

○   The standard deviation of $5,000 at the Sky Corporation indicates that the salaries of the employees are $5,000 above the mean of $70,000.

○   The standard deviation of $5,000 at the Sky Corporation indicates that the salaries of the employees are $5,000 below the mean of $70,000.

○   The standard deviation of $5,000 at the Sky Corporation indicates that the salaries of the employees are $5,000 more than the mean of $70,000.

Question 14

14. Interpret the standard deviation of $50,000.

○   The standard deviation of $50,000 at the Earth Corporation indicates that the salaries of the employees are typically $50 away from the mean of $70,000.

○   The standard deviation of $50,000 at the Earth Corporation indicates that the salaries of the employees are typically $50,000 above the mean of $70,000.

○   The standard deviation of $50,000 at the Earth Corporation indicates that the salaries of the employees are typically $50,000 below the mean of $70,000.

⦿   The standard deviation of $50,000 at the Earth Corporation indicates that the salaries of the employees are typically $50,000 away from the mean of $70,000.

Question 15

15. Compare the two standard deviations; what do they tell us about the salaries of the employees of the two corporations? More specifically, which of the following statements is true?

○ The typical salaries at the Sky Corporation are closer to the mean and the typical salaries at the Earth Corporation are further away from the mean.

○ The typical salaries at the Sky Corporation are lower than the average and the typical salaries at the Earth Corporation are higher than the average.

○ There is no difference in the salary variations of the two corporations.

○ The average salaries at the Sky Corporation are higher than the average salaries at the Earth Corporation.

○ The average salaries at the Sky Corporation are lower than the average salaries at the Earth Corporation.

Question 16

16. Compare the two standard deviations; what do they tell us about the salaries of the employees of the two corporations? More specifically, which of the following statements is true?

○ The salaries of the Sky Corporation are more similar (more homogeneous) and the salaries of the Earth Corporation are more dissimilar (more heterogeneous).

○ Employees at the Sky Corporation make more money than their counterparts at the Earth Corporation.

○ There is no difference in the salary variations of the two corporations.

○ The average salaries at the Sky Corporation are higher than the average salaries at the Earth Corporation.

○ The average salaries at the Sky Corporation are lower than the average salaries at the Earth Corporation.

Question 17

Questions 17 - 20. A claims adjustment manager learns that her 200 employees (N = 200) spend an average of 6 hours per claims case with a standard deviation of 2 hours. The numbers of hours are normally distributed.

17. What is the probability of finding an employee who spends more than 10 hours per claims case?

Note: This answer requires an exact answer and no partial credit will be given. The answer must be a number that has four digits after the decimal. For example: .1234 or 0.1234.

Hint: We need to use a two step process. First, we calculate a z score using the formula: Z = (Specific Score - Mean) / Standard deviation. (10-6)/2 = 4/2 = 2. This means that 10 hours are 2 standard deviations away from the mean of 6. So our z score is 2. Second, we get the probability from the table on Page 95 using the obtained z score.

Answer | .0228

Question 18

18. What is the probability of finding an employee who spends more than 8 hours per claims case? Note: this answer requires a number that has four digits after the decimal. For example: .1234, or 0.1234. No partial credit will be given if the answer does not conform to this format. Hint: Use a two step process. First get a z score and then get the probability from the table on Page 95.

Answer | .1587

Question 19

19. How many employees spend more than 8 hours per claims case? Note: The answer is a number that has two digits after the decimal. For example: 12.12. Do not round off to a whole number. No partial credit will be given to answers that do not conform to this format. Hint: In

order to get the number of cases, we multiply the obtained probability by the total number of cases. The total number of cases is 200.

Answer | 31.74

Question 20

20. What is the probability of finding an employee who spends more than 14 hours per claims case? Hint: We need to take a two step process. First we calculate the z score that goes with 14, using the formula: (specific score - mean)/standard deviation. $(14-6)/2 = 8/2 = 4$. Then we go to Page 95 to get the probability.

- ● Less than .0001
- ○ .1587
- ○ .0012
- ○ .0228

Question 21

21-25. The City of Hills has 10,000 residents. They plan to spend an average of $1,000 on gifts this year. The standard deviation is $200. Assume that the gift amounts are normally distributed.

21. We want to know the probability of spending more than $1,600. As a first step, we calculate the z score for $1,600. What is that z score? Note: It is a one digit number without any decimals. For example: 9.

Hint: We get the z score by using the z score formula: (specific score-mean)/standard deviation. The specific score is $1,600, the mean is $1,000, and the standard deviation is $200.

Answer | 3

Question 22

22. What is the probability of spending more than 1,600 dollars? It is a number with 4 digits after the decimal point. For example: .0123 or 0.0123. Hint: We can use the z score that we obtained in the previous question and get the probability from the table on the textbook Page 95.

Answer .0014

Question 23

23. What is the probability of spending less than 400 dollars? Hint: We need to take a two step approach. First, we get the z score that goes with 400 dollars by using the z score formula: (specific - mean)/ standard deviation. The specific score is 400 dollars. The mean is 1,000 dollars. The standard deviation is 200 dollars. Once we get the z score, we go to Page 95 and get the probability.

- ⦿   .0014

- ○   22.8

- ○   .1587

- ○   .5000

Question 24

24. What percentage of the residents will spend more than 1,400 dollars? Hint: We need to take a three step approach. First, we get the z score with the formula: (specific score - mean)/ standard deviation. The specific score is 1,400 dollars. The mean is $1,000. The standard deviation is $200. Then we go to the table on Page 95 and get the probability. Finally, we multiply that probability by 100 to convert that into a percentage.

- ○   .49%

- ⦿   2.28%

○   .14%

○   15.87%

Question 25

25. How many residents will spend more than $1,400? Note: The answer is a 3 digit number without any decimal. For example: 123. Hint: Multiply the probability (not the percentage) obtained in the previous question by the total number of cases, which is 10,000.

Answer [228]

Question 26

25. Identify the cases, independent variable, and dependent variable in the following hypothesis: Employees with more spirituality are less likely to be absent from their work than employees with less spirituality.

○   Cases: Likelihood to be absent from work; Independent variable: Levels of spirituality; Dependent variable: More absence

○   Cases: Spirituality; Independent variable: Levels of spirituality; Dependent variable: less absence

◉   Cases: Employees; Independent variable: Levels of spirituality; Dependent variable: Likelihood to be absent from work

○   Cases: Employees; Independent variable: Likelihood to be absent from work; Dependent variable: more absence

Sample Final Exam Questions

Question 1

Questions 1 - 4. An analyst has a research hypothesis: Employees who use more technology in the process of performing their jobs are more confident about their job competence than employees who use less technology in the process of performing their jobs.

1. State the null hypothesis.

○   A. Employees who use more technology in the process of performing their jobs are less confident about their job competence than employees who use less technology in the process of performing their jobs.

◉   B. Employees who use more technology in the process of performing their jobs are not more confident about their job competence than employees who use less technology in the process of performing their jobs.

○   C. Employees who use more technology in the process of performing their jobs are much less confident about their job competence than employees who use less technology in the process of performing their jobs.

○   D. Employees who use more technology in the process of performing their jobs and employees who use less technology in the process of performing their jobs are equally confident about their job competence.

Question 2

2. What are the cases?

○   A. Patients

◉   B. Employees

○   C. Managers

○   D. Supervisors

Question 3

3. What is the independent variable?

○   A. Job competence

○   B. Levels of job competence

◉   C. Levels of technology use

○   D. Employees

Question 4

4. What is the dependent variable?

○   A. Technology Use

○   B. Job competence

○   C. Employees

◉   D. Levels of confidence about job competence

Question 5

Questions 5 - 6. A marketing researcher learns that young adults in his city purchased an average of 20 DVD movies with a standard deviation of 5 DVD movies last year. He obtained this information from a survey of 500 young adults. Assume that the numbers of DVD movies purchased are normally distributed.

5. What is the probability of a young adult purchasing more than 30 DVD movies last year? Note: This question can be answered by going through a two-step process. First get the z score using the formula: $Z = $ (Specific Score - Mean) / Standard Deviation. The specific score is 30, the mean is 20, and the standard deviation is 5. Then get the probability from the probability table on page 95.

○   A. -.0228

○   B. .1587

⊙   C. .0228

○   D. .0014

Question 6

6. How many young adults purchased less than 15 DVD movies last year? Note: This question can be answered by going through a three step process. First get the z score using the formula: Z = (Specific Score - Mean) / Standard deviation. Then get the probability from the probability table on page 95. Finally, multiply the obtained probability by the total number of cases. The total number of cases is 500.

○   A. 22.80

⊙   B. 79.35

○   C. 500

○   D. 11.40

Question 7

Questions 7-10. A restaurant manager wants to learn which style of guestroom decoration is more appealing to her guests. She decorates one guestroom like a bright garden and another guestroom like a shady garden. Based on her marketing research, she suspects that the bright garden room is more attractive to her guests than the shady garden room is. To confirm her suspicion, she counts every day the number of guests who request each room for two months. Can she conclude that the bright garden room is more popular than the shady garden room?

Number of guests by type of guestroom decoration

Decoration type	Mean	Standard Deviation	# of Cases
Bright garden	100	10	60
Shady garden	85	12	60

7. The research hypothesis is:

○ A. The bright garden room is more attractive than the shady garden room.

○ B. The bright garden room is not more attractive than the shady garden room.

○ C. The bright garden room is less attractive than the shady garden room.

○ D. The shady garden room is more attractive than the bright garden room.

Question 8

8. The null hypothesis is:

○ A. The bright garden room is more attractive than the shady garden room.

○ B. The bright garden room is less attractive than the shady garden room.

○ C. The bright garden room is not more attractive than the shady garden room.

○ D. The shady garden room is more attractive than the bright garden room.

Question 9

9. The independent variable is:

○ A. Garden room

○ B. Shady room

○ C. Type of guestroom (managers', employees')

○ D. Type of guest room (bright garden, shady garden)

Question 10

10. The dependent variable is:

○   A. Guest rooms

◉   B. Attractiveness or popularity of guestroom decoration measured by the number of guests

○   C. Managers

○   D. Bright garden room

Question 11

Questions 11-16. Newspapers report that new internet companies in the East are more likely to fail than new internet companies in the West. A reporter agrees with that claim and wants to confirm it. She learns that 40 of the 100 internet companies in the East and 10 of the 100 internet companies in the West failed within a year of operation. After constructing a crosstabulation, calculate and interpret column percentages.

11. What is the independent variable?

○   A. Newspapers

○   B. Reporters

◉   C. Region (East, West)

○   D. Internet companies

Question 12

12. What is the dependent variable?

○   A. Newspapers

◉   B. Failure Status (failed, not failed)

○   C. Region (East, West)

○   D. Amount of revenue

Question 13

13. The frequencies for Cell1, Cell2, Cell3, and Cell4 are:

○   A. 90, 60, 10, 40

◉   B. 40, 10, 60, 90

○   C. 10, 40, 90, 60

○   D. 60, 90, 10, 40

Question 14

14. The column percentages for Cell1, Cell2, Cell3, and Cell4 are:

◉   A. 40, 10, 60, 90

○   B. 10, 40, 90, 60

○   C. 90, 60, 10, 40

○   D. 10, 40, 90, 60

Question 15

15. Calculate a column percentage and interpret it.

○   A. The new internet companies in the East are 20% more likely to fail than the new internet companies in the West.

◉   B. The new internet companies in the East are 30% more likely to fail than the new internet companies in the West.

○   C. The new internet companies in the West are 20% more likely to fail than the new internet companies in the East.

○   D. The new internet companies in the East are 40% more likely to fail than the new internet companies in the West.

Question 16

Questions 16 - 19. Work with the following regression equation.

Amount of net profit (in dollars) = -50 + 40 Number of items sold.

N = 60   R-square = .95

16. The independent variable and the dependent variable in the equation are:

- ○ A. Independent variable: Amount of sales; Dependent variable: Number of items sold
- ○ B. Independent variable: Companies; Dependent variable: Number of items sold
- ◉ C. Independent variable: Number of items sold; Dependent variable: Amount of net profit
- ○ D. Independent variable: Amount of debt; Dependent variable: Amount of net profit

Question 17

17. Interpret the regression coefficient.

- ○ A. When the number of items sold goes up by 1, the amount of net profits goes up by $$80.
- ○ B. When the amount of net profits goes up by 1$, the number of items sold goes up by $40.
- ◉ C. When the number of items sold goes up by 1, the amount of net profits goes up by $40.
- ○ D. When the number of items sold is 1, the amount of net profits is $40.

Question 18

18. Interpret the Y-intercept.

○   A. When the number of items sold is 0, the amount of net profits is 50 dollars.

○   B. When the number of items sold is 1, the amount of net profits is 50 dollars.

◉   C. When the number of items sold is 0, the amount of net profits is -50 dollars.

○   D. When the number of items sold is 0, the amount of net profits is $40.

Question 19

19. When the number of items sold is 5, what is the predicted amount of net profit?

◉   A. It is $150. It is obtained by: -50 + 40 (5) = 150.

○   B. It is $30. It is obtained by: -50 + 40(2) = 30.

○   C. It is $-50. It is obtained by: -50 + 40(0) = -50.

○   D. It is $-10. It is obtained by: -50 + 40(1) = -10.

Question 20

Question 20-23. A human resources manager thinks that employees who work in a more intellectually stimulating work environment are more satisfied with their jobs than those who work in a less intellectually stimulating work environment. She selects a random sample of 400 employees and measures their work environment (high-more stimulating, low-less stimulating) and job satisfaction (high, low). He learns that 150 of the 200 employees who work in a more stimulating work environment report high job satisfaction and that 100 of the 200 employees who work in a less stimulating work environment report low job satisfaction. He wants to know whether he should make the work environment more stimulating.

Note 1: X = Independent variable. Y = Dependent variable.

Note 2: L= Low, H = High.

20. Which of the following hypothetical data is appropriate for addressing this management challenge?

Hint: The independent variable and the dependent variable are both measured at the ordinal level. Therefore, the means and standard deviations (which require interval-ratio level data) cannot be calculated and they do not exist.

Note: Levels of measurement (Chapter 2), Difference of means test (Chapter 8), crosstabulation analysis (Chapter 10), and regression analysis (Chapter 13) should be reviewed before answering these questions. Appendix D has some relevant information as well. In general, regression analysis requires interval-ratio independent variables and interval-ratio dependent variables. Crosstabulation analysis uses nominal or ordinal independent variables and nominal or ordinal dependent variables. Difference of means tests use nominal or ordinal independent variables and interval-ratio dependent variables.

A X = 1,30,21,40,32,40,35, ... Y = 3,2,4,5,6,7,3,4,5,6,7,8,.......
.   Which can be reduced to: Y = 100 +150X

B X = H,L,L,L,H,H,L,H,H,H,...... Y = H,L,L,H,H,L,H,,L,L,H,.....
.   Which can be summarized as:

	Mean	Standard deviation
More Stimulating Work Environment (H)	150	8
Less Stimulating Work Environment (L)	80	10

C X = H,L,L,L,H,H,L,H,H,H,...... Y = H,L,L,H,H,L,H,L,L,H,.....
.   Which can be tabled as:

		Stimulus Levels of Work Environment (X)	
		L(Less stimulating)	H (More stimulating)
Job Satisfaction (Y)	L	100	50
	H	100	150

D X = H,L,L,L,H,H,L,H,H,H,...... Y = H,L,L,H,H,L,H,,L,L,H,.....
.   Which can be summarized as: Average Job Satisfaction Rate

	Mean	Standard deviation
More Stimulating Work Environment (H)	150	8
Less Stimulating Work Environment (L)	100	10

Question 21

21. Analyze the data that you chose in the previous question and interpret the outcome. Which of the following is the appropriate interpretation of the analysis outcome?

- ⦿ A. Those who work in a more stimulating work environment are 25% more likely to be satisfied with their jobs than those who work in a less stimulating work environment.

- ○ B. We conclude that the average job satisfaction score of those who work in a less stimulating work environment is significantly higher than those who work in a more stimulating work environment. When we claim this, we would be wrong less than 5% of the time.

- ○ C. When the work environment score goes up by 10, the job satisfaction score goes up by 20.

- ○ D. Those who work in a more stimulating work environment are 75% more likely to be satisfied with their jobs than those who work in a less stimulating work environment.

Question 22

22. Given the outcome of the analysis, what would you recommend to the human resources manager?

- ○ A. Try to do nothing because the outcome is inconclusive.

- ○ B. Try to make the work environment less stimulating so that job satisfaction can go up.

- ○ C. Try to do nothing because the data are not analyzed appropriately.

- ⦿ D. Try to make the work environment more stimulating so that job satisfaction can go up, particularly if there are no other factors that can account for job satisfaction.

Question 23

23. Variables can be measured at several different levels. The level of measurement determines or constrains the types of statistical method

that can be used to address a given management challenge. Which of the following hypothetical statements would be correct?

   A. If we measure the work environment variable at the nominal level and the job satisfaction variable at the interval-ratio level (with an index of job satisfaction), we can use crosstabulation analysis to examine the relationship between these two variables.

   B. If we measure the work environment variable at the ordinal level (with two categories: more stimulating work environment and less stimulating work environment) and the job satisfaction variable at the ordinal level (with two categories: more satisfaction and less satisfaction), we can do regression analysis to examine the relationship between these two variables.

   C. If we measure the work environment variable at the ordinal level and the job satisfaction variable at the interval-ratio level (with an index), we can do a difference of means test to examine the relationship between these two variables.

   D. If we measure the work environment variable at the interval-ratio level (with an index) and the job satisfaction variable at the ordinal level, we can do regression analysis to examine the relationship between these two variables.

Question 24

24-27. A restaurant manager suspects that servers who smile more often (high level) receive a higher amount of tips than servers who smile less often (low level). In order to test her suspicion, she collects the following information:

Amount of daily tips in dollars by the levels of smiling

Level of smiling	Mean	Standard deviation	Number of cases
High	300	50	50
Low	200	100	50

24. What are the independent variable and the dependent variable?

○   A. Independent variable: Manager; Dependent variable: Amount of tips

○   B. Independent variable: Amount tips; Dependent variable: Levels of smiling

○   C. Independent variable: Small amount of tips; Dependent variable: Median tips

◉   D. Independent variable: Levels of smiling; Dependent variable: Amount of tips

Question 25

25. At what level are the independent variable and the dependent variable measured?

○   A. Independent variable: Ordinal; Dependent variable: Ordinal

◉   B. Independent variable: Ordinal; Dependent variable: Interval-ratio

○   C. Independent variable: Interval-ratio; Dependent variable: Nominal

○   D. Independent variable: Ordinal; Dependent variable: Nominal

Question 26

26. What statistical method would you use to address this management challenge?

○   A.   Standard deviation

○   B.   Crosstabulation analysis

◉   C.   Difference of means test

○   D.   Regression

Question 27

27. Variables can be measured at different levels. The level of measurement determines or constrains the types of statistical method that can be used to address management challenges. Which of the following hypothetical statements would be correct?

- ◉ A. If we measure the levels of smile variable at the ordinal level (low, high) and the amount of tips variable at the ordinal level (low, high), we can use crosstabulation analysis to examine the relationship between these two variables.

- ○ B. If we measure the levels of smile variable at the interval-ratio level (number of smiles) and the amount of tips variable at the ordinal level (low, high), we can use crosstabulation analysis to examine the relationship between these two variables.

- ○ C. If we measure the level of smile variable at the ordinal level nominal level (low, high) and the amount of tips variable at the ordinal level (low, high), we can use regression analysis to examine the relationship between these two variables.

- ○ D. If we measure the levels of smile variable at the interval-ratio level (number of smiles) and the amount of tips variable at the ordinal level (low, high), we can use a difference of means test to examine the relationship between these two variables.

Made in the USA
Charleston, SC
03 January 2014